TEXAS ROOTS

Number Eight:
TEXAS A&M UNIVERSITY AGRICULTURE SERIES

C. ALLAN JONES

Texas Roots

Agriculture

AND

Rural Life

BEFORE THE

Civil War

Texas A&M University Press
COLLEGE STATION

First edition

The paper used in this book meets the minimum requirements of the American National Standard for Permanence of Paper for Printed Library Materials, Z39.48-1984.
Binding materials have been chosen for durability.
∞

Library of Congrress Cataloging-in-Publication Data

Jones, C. Allan.
Texas roots : agriculture and rural life before the Civil War / C. Allan Jones.—1st ed.
p. cm.— (Texas A&M University agriculture series ; no. 8)
Includes bibliographical references and index.
ISBN 1-58544-418-9 (cloth : alk. paper) —
ISBN 1-58544-429-4 (pbk. : alk. paper)
1. Texas—History—To 1846. 2. Texas—History—1846–1950. 3. Texas—Social life and customs. 4. Ranch life—Texas—History. 5. Farm life—Texas—History. 6. Plantation life—Texas—History. 7. Agriculture—Texas—History. 8. Agriculture—Social aspects—Texas—History. I. Title. II. Series.
F389.J66 2005
976.4'01—dc22

2004018516

To Jef, Erin, and Anna.
With special thanks to Rosemary Payton, Clint Wolfe, Mary Lenn Dixon, Shannon Davies, Jennifer Ann Hobson, Jennifer Jacobs, and Michal Bagley.

CONTENTS

TEXAS ROOTS

INTRODUCTION

Texas is a land of great natural beauty, enormous human and natural resources, and proud people. Texans have been marked by their state's Hispanic heritage, the war for independence from Mexico, the state's survival as an independent republic, and a frontier ethic defined by over two hundred years of conflict among Native American, Hispanic, Anglo-American, and other ethnic groups. Because of their state's colorful history, great size, and fruitful yet sometimes harsh landscape, Texans have maintained a strong, personal relationship with both their state and its lands. That relationship often begins in childhood, is cultivated in the public schools, and matures in adulthood. As a result, many Texans have a strong sense of place and take great pride in their state and in being Texan.

The objective of this book is to bring alive a part of Texas history that is rarely addressed: the relationship of Texans to their land before the Civil War—the time when the foundations of Texans' identity were forged. This approach focuses on how Texans used the land, including hunting, cultivating crops, producing livestock, and constructing homesteads—activities that occupied, directly or indirectly, the great majority of Texans prior to the Civil War. I have used numerous quotations from primary sources to give the reader a better sense of Texas and Texans in the eighteenth and nineteenth centuries. These quotations often reveal quite clearly the levels of education as well as the attitudes of their speakers. As much as possible, I have left the quotations as I have found them, though it has often been convenient to use selected parts of longer sentences to incorporate them into the narrative. If nothing else, I hope that this work will stimulate the reader to search out and read the original sources. Fortunately, many of them are in print and readily available.

To help focus the work, I have chosen to emphasize the two principal agricultural traditions in Texas from the beginning of its recorded history until the Civil War: the Hispanic Tejanos and the English-speaking Texians. Through these two peoples and their interactions, we can better understand the foundations of Texas' post–Civil War economy, as well as Texans' connections to their ethnic roots and the land. But why were agriculture and rural life prior to 1860 important to the development of Texas? First, it was the period when the agricultural systems and lifestyles of New Spain and the southeastern United States were adapted to the Texas environment. Second, it was the period when Spanish and Mexican traditions and institutions came into contact with their Anglo-American counterparts, often producing particularly Texan adaptations of these technologies, laws, and customs.

Tejanos is a term that has been used since at least the 1820s to refer to the Hispanic residents of Texas. The Spaniards initially settled in Texas in the late seventeenth and early eighteenth centuries to convert the Caddos and other Indians to Catholicism, as well as to prevent French incursions into lands claimed by Spain. Seeking to establish a presence between eastern Texas and the closest Spanish settlements on the Río Grande, missionaries, soldiers, and private farmers and ranchers began to settle and develop the valley of the Río San Antonio beginning in about 1720. Conflicts among missionaries, soldiers, private citizens, Indians, and the Crown's administrators limited the region's development. However, the ranching industry that grew up between the Río San Antonio and Río Grande after about 1750 became a key component of the Texas economy for more than a century.

The term *Texian* was coined in the early 1800s and referred to English-speaking immigrants, most of whom came to Texas from the southern United States. The Texians began to arrive in substantial numbers after Mexico gained its independence from Spain in 1821. In the 1820s and early 1830s the Tejanos accepted them as loyal Mexican subjects. The Texians adapted southern agricultural traditions to the Texas environment and for more than a decade lived under Mexican rule. But religious, cultural, and economic issues brought political divisions, leading in

1836 to the Texas Revolution and formation of the Republic of Texas. Despite financial panic and continuing conflicts with Mexicans and Indians, Texians expanded agriculture and ranching into the northern and western parts of central Texas. During the 1840s and 1850s the number of farms and farmers increased. Slave owners expanded their operations, and herds and crop production continued to grow. This rapid agricultural development stimulated the growth of towns, laying the foundations of post–Civil War expansion of agriculture and manufacturing. After the Civil War the term *Texian* was generally replaced by *Texan.*

The terms *Tejano, Texian,* and *Texan* are all derived from the Caddo term *techas,* usually translated as "friend." The westernmost of the great Mississippian nations, the Caddos' highly productive corn-based agriculture, complemented by extensive hunting and gathering, permitted them to dominate northeastern Texas for hundreds of years. However, by the time the Texians arrived in the early nineteenth century, Caddo communities had been decimated by European diseases. As a result, the Caddos left little lasting impact on Texian agriculture.

Now, a few words about sources. In recent years a number of excellent books have been written about the history of early Texas. The work of historians has helped me, an agricultural scientist without formal historical training, to understand much of the social and political context in which agriculture developed. Excellent works by Armando Alonzo, Andrés Tijerina, Terry Jordan, Jack Jackson, Richard Lowe, David La Vere, Timothy Perttula, Abigail Curlee Holbrook, Randolph Campbell, and others have provided both the details and context I needed for this work. In addition, the Béxar Archives translations and many early accounts that have been reprinted or published for the first time are rich sources of primary information. Some of the most interesting and informative accounts of these topics were written by travelers such as Pierre Pagés, Jean Louis Berlandier, Noah Smithwick, and Frederick Law Olmsted. It has been a great pleasure to piece together information from such diverse materials into a coherent, if inevitably incomplete, picture of early Texas agriculture and rural life.

Finally, in order to focus on agriculture and rural life, I have minimized discussions of Texas' rich political and social history. Readers may want to consult a general history such as Randolph Campbell's *Gone to Texas: A History of the Lone Star State* to supplement this more limited treatment.

Part I

LOS TEJANOS

FARMING AND RANCHING IN HISPANIC SOUTH TEXAS

Chapter 1

INDIAN AND SPANISH COLONIAL ORIGINS

NOWHERE IS THE HISPANIC COMPONENT OF Texas culture more important than between the Río San Antonio and the Río Grande. The agriculture of Hispanic south Texas grew out of the interactions of very different traditions. Its roots were the ancient corn-based systems of pre-Columbian Mexico that were enriched beginning in the early sixteenth century by incorporation of the wheat- and livestock-based agriculture of Spain. This chapter describes how the hybrid farming and ranching culture of Hispanic Texas developed from these two rich traditions, first supporting the Spanish missions along the Río San Antonio, then evolving into the Tejano ranching culture of south Texas.

PRE-COLUMBIAN AGRICULTURE

When the Spaniards arrived in Mexico in the early sixteenth century, they encountered a well-developed agriculture based, in almost all areas, on cultivation of corn, beans, squash, chilies, and a wide array of fruits, vegetables, herbs, and medicinal plants. The systems used to grow these crops varied greatly from place to place, ranging from the commonly used "slash-and-burn" process to more intensive and elaborate systems. Slash-and-burn systems were common throughout much of the Americas, especially on steep hillsides and other lands with shallow, infertile soils. Indian farmers cut small plots of forest or woodland, usually no more than a few acres at a time, with stone or copper axes and left the plant residues to dry on the ground. Shortly before the beginning of the rainy season they burned the residues, which softened and sterilized the soil, killed insects, and left behind a layer of ash rich in potassium and phosphorus. As soon as the wet season began, farmers planted seeds directly into the ashes, using a *coa,* or digging stick, to make small holes and dropping three to six seeds into each one. When necessary, they also cut or uprooted weeds with a coa.[1]

Indian farmers grew many varieties of corn, beans, squash, and chilies, selecting those best adapted to their regional environments and preferences. They planted the crops together, often a few weeks apart, with climbing beans using the corn stalks for support. Farmers bent the corn stalks over just below the ear after the grains were filled but before the plants were dry. This reduced damage by birds and rain, and it allowed the ears to be harvested a few at a time over several months. Corn-grain yields of fifteen to fifty bushels per acre were probably common, with the first crop after the land was cleared normally yielding more than later crops. But after a few seasons, soil fertility declined, weeds became more difficult to control, and farmers abandoned the land for several years while the natural vegetation regrew and soil fertility returned. The Indians also cultivated a wide variety of other vegetables, fruits, and herbs, both in small gardens near their houses and in their fields. These crops included tree fruits, such as custard apples, guavas, sweet sapotas, and avocados, as well as tomatoes and prickly pear pads and fruits. Slash-and-burn systems were common throughout Mexico and Central America, and there must have been dozens of variations on the same theme, each an adaptation to the local environment and needs. Similar farming systems developed in the woodlands of eastern North America. In eastern Texas, the Caddos developed sustainable corn-based, slash-and-burn agriculture, which they combined with extensive hunting and gathering to build a complex and powerful confederacy.[2]

Some farmers practiced more intensive agriculture, especially in areas with high population densities. In dry climates the Indians built small structures of stones, wood, or soil to slow and direct runoff to plots where it could infiltrate and later be used by plants. On steep slopes farmers constructed impressive systems of parallel stone terraces to slow runoff, increase infiltration of water, and provide narrow strips of deep soil on which crops could be grown. Beginning about three thousand years ago in Oaxaca and the Valley of Mexico, communities developed a variety of irrigation sources and methods. Major canal networks irrigated areas of up to twenty-four hundred acres, and well before the arrival of the Spaniards most of the lands in the Valley of Mexico were producing irrigated crops. *Chinampas,* representing the most intensive agricultural system developed in Mexico, were built in and around large, shallow lakes such as Texcoco and Xochimilco. They consisted of canals surrounding long, rectangular plots or beds that the farmers built up with aquatic plants and soils excavated from the canals. This rich silt and organic matter maintained soil fertility, and the porous soil of the beds allowed water from the canals to infiltrate and subirrigate the crops, producing an abundant and varied food supply.[3]

SPANISH COLONIAL AGRICULTURE

The Spanish conquistadores and clerics came to the Americas to gain power, fortunes, and souls. They did not come to live within the indigenous cultures, but to dominate, convert, and rule them. This mentality, operating in conjunction with the decimation of Indian peoples by European diseases, produced dramatic changes in food production. Spaniards introduced horses, burros, mules, and cattle, classified as large livestock, or *ganado mayor,* and small animals such as sheep, goats, and hogs, known as *ganado menor.* Horses increased the speed of communication. Oxen, burros, and mules pulled plows and carts. Cattle, swine, sheep, and goats provided much-needed animal protein, utilizing steep grasslands and woodlands that the Indians, without large herbivores, had not been able to exploit. Sheep also produced wool for clothing. Spaniards brought wheat, barley, rice, and sugarcane as well. Fruits included grapes, olives, oranges, lemons, walnuts, apples, pears, figs, peaches, watermelons, mangoes, and plantains. Old-world vegetables and herbs included lettuce, cabbage, collards, cucumbers, chickpeas, fava beans, radishes, onions, mint, rue, cilantro, and parsley. Both the Spaniards, especially the clergy, and the Indians began to select those varieties that were able to grow and produce the best fruit in the wide variety of environments of the New World.[4]

During the sixteenth and seventeenth centuries the Spaniards brought a number of European agricultural technologies to the New World. They replaced pre-Hispanic copper hatchets with iron axes. Machetes with steel blades replaced similar wooden tools that had cutting edges made of obsidian. Ox-drawn plows were introduced, and European-style hoes replaced Indian coas. Indian farmers began to use draft animals, including horses, burros, mules, and oxen, to carry goods and pull carts. Water- and animal-powered mills began to grind wheat, corn, and sugarcane. Indian leaders, or *caciques,* adopted Spanish dress and introduced European tools and crops into the communities under their control. Soon Spanish merchants were encouraging the production of *cochineal* (a bloodred dye made from insects collected from prickly pear cactus), woolen blankets, tobacco, and silk. These products, as well as a wide variety of both indigenous and European foodstuffs, began to move throughout Mexico on the backs of animals and in carts.[5]

Though yields of wheat were less than those of corn, wheat could be produced with much less labor, a real advantage as European diseases decimated the Indian labor force in the sixteenth century. The Spanish farmers sowed wheat by broadcasting the seeds over the field, then lightly harrowing the surface to cover them. The closely spaced plants crowded out most of the weeds, further reducing labor requirements. By the 1560s hundreds of farmers were raising wheat in well-watered valleys near the major cities. Eventually, producers included not only owner-operators who lived on the farm or in a nearby village but also sharecroppers, tenant farmers, and managers working for absentee landowners. During the sixteenth century they established municipal granaries to regulate the amount of wheat on the retail market, thereby stabilizing prices.[6]

Because Spaniards considered plowing, planting, and harvesting beneath their dignity, Indians provided most of the labor. During the first years after the conquest, many Indian communities were forced to pay Spanish *encomenderos,* conquistadores with the right to demand tribute from these communities in exchange for protection and religious training. In an effort to reduce abuses, the Spaniards introduced the *repartimiento* system in 1549. It required that Indian communities provide labor to Spanish mines and farms, but under the supervision of officers of the Crown rather than the owners or managers themselves. By 1632, in response to declining Indian populations, most repartimientos and other forms of forced labor were abolished.[7]

The Spaniards had developed large-scale cattle ranching practices during the Middle Ages on the dry interior tablelands of the Meseta Central. There they began acquiring the techniques that they would later take to the New World, including the use of horses, the open range, roundups, branding, and long-distance drives. Herds were managed by a foreman (called a *mayoral* or *mayordomo*) and several cowboys (or *vaqueros*), who were either bonded servants or freemen, usually contracted annually and paid in cash or livestock. They rode small, hardy horses developed in Andalusía and by the late fifteenth century had adopted light Moorish saddles with short stirrups. Dogs assisted in guarding and managing the cattle. Another important innovation, the stock-raisers' guild, or *mesta,* was established to regulate the cattle-ranching industry. It assigned grazing rights, set wages of *vaqueros,* regulated the use of brands, set the rate of compensation for crops damaged by livestock, and supervised the roundups, or *rodeos,* that were held in the spring and autumn to brand calves and return strays to their rightful owners. It also regulated the marketing and sale of cattle in town markets, the slaughter of animals, and the sale of meat in butcher shops and town fairs, or *ferias;* mediated disputes; and imposed penalties for illegal activities, including theft, brand changing, and killing another owner's stock. The principal breeds of Spanish cattle included black *ganado prieto,* the reddish or tan *retinto,* and the *barrenda,* which were predominantly white with black markings. The sheep raised were the *chaurro,* lean and hardy animals weighing sixty to eighty pounds at maturity and yielding one to two and a half pounds of coarse wool. The Spaniards' tough, adaptable goats were used primarily for meat, often consumed as *cabrito,* or roasted kid.[8]

The Mexican cattle industry began around Tampico and Veracruz on the Gulf of Mexico and on the Mesa Central near Mexico City. In the late 1530s and 1540s cattle numbers increased rapidly, and the government encouraged ranchers to move northwest of Mexico City to El Bajío, a region that stretched from Guadalajara in the west to Querétaro in the east and provided lush pastures of perennial grasses. But as cattle ranching moved northward, it began to suffer losses from poor, displaced Indians who stole cattle for meat, tallow, and hides. As a result, in 1537 the first mesta was established in Mexico, requiring that owners of more than three hundred ganado menor or twenty ganado mayor become members.[9]

Stimulated by the discovery of silver in Zacatecas and other locations in northern Mexico in the late 1540s, the cattle industry spilled northward onto the north-central plateau, or Mesa del Norte, bringing it into conflict with the fierce nomadic tribes of the region, known collectively as *Chichimecas.* From 1550 to 1585 Spanish authorities waged a concerted war with these tribes. But beginning in the 1590s government policy changed from warfare to religious conversion and assimilation. As a result, the church began to have a more active role along the northern frontier. In order to support their populations of Indian residents, the missions developed herds of livestock. *Presidios,* garrisons charged with protecting the missions, maintained large herds of horses to serve as mounts, and private ranches provided meat for soldiers and civilians. The government, in order to gain some level of control over the expanding livestock industry, began to award *estancias,* or large grants of land, in unsettled areas at a distance from Indian villages. During the late

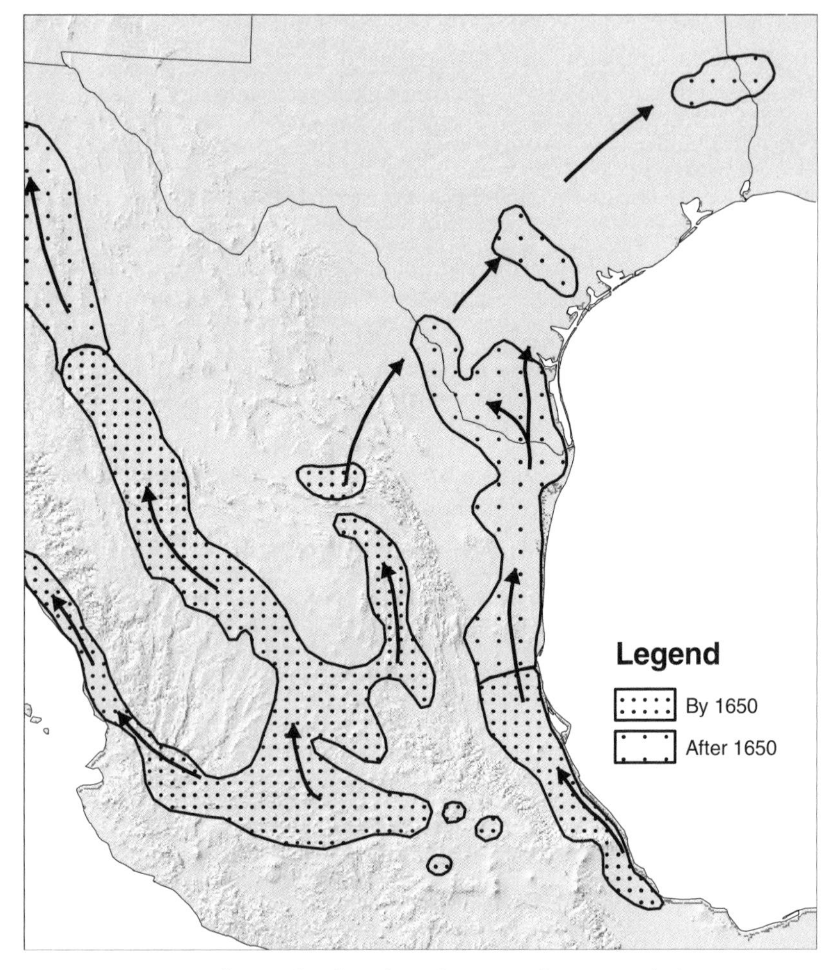

MAP 1. *Expansion of Spanish colonial ranching. Ranches present before 1650 indicated with heavy stippling, those added after 1650 indicated with light stippling.*

sixteenth century the great ranches, or *haciendas,* of northern Mexico developed, extending their dominion over hundreds of thousands of acres. But overgrazing began to degrade the fragile rangeland, ranchers killed large numbers of cattle for their hides and tallow, and thieves and packs of wild dogs also took an increasing toll. By the late sixteenth and early seventeenth centuries cattle numbers had declined, causing meat and cattle prices to increase. Able to thrive on poorer pastures than cattle, sheep became a more important component of many ranches, especially in the highlands of northern Mexico.[10]

By 1650 cattle ranching occupied a large region from the Mesa Central northward through El Bajío, along the moist eastern flank of the Sierra Madre Occidental and to a lesser extent on the western slopes of the Sierra Madre Oriental. Cattle ranching also expanded along the eastern coast between Tampico and Veracruz; and along the western coast northwest of Guadalajara. After 1650 the industry expanded farther northward along the coasts of the Gulf of Mexico and Gulf of California, into the region to the west and northwest of Chihuahua City, and into the valleys of the northern Sierra Madre Occidental. The cattle raisers avoided steep mountain slopes; the Gran Tunal, a large area north of San Luis Potosí dominated by prickly pear cactus and deciduous shrubs; and the even drier Chihuahuan desert, stretching northwestward between the two sierras. The ranchers in the highlands often raised many more sheep than cattle. Their flocks were guarded by *pastores,* or sheepherders, with the help of dogs. In contrast, the ranchers in the grasslands along the Gulf of Mexico north of Veracruz specialized in raising cattle and horses.[11]

The large livestock-based haciendas on the northern frontier took on the characteristics of feudal estates. The owner, or *hacendado,* presided over a largely self-sufficient empire with livestock, crops, workshops, stores, a church, and almost anything else its inhabitants might need. Meat was consumed at the hacienda or sold in the cities or the mines. Hides were sent to dealers for export to Spain, for use in the mines, or for processing into leather. Sheep and goats were other important sources of meat, and wool was sold to mills in the major cities to be woven into cloth, most of which was consumed in the Americas.[12]

By the late sixteenth century the Mexican vaqueros had adopted the basic clothing, equipment, and methods that would be taken to Texas over a century later. Most wore a cotton or wool *camisa* (shirt), a leather *chaqueta* (jacket), tight knee-length *sotas* (pants) laced up the sides, leather *botas* (leggings) wound around the legs below the knees, and a simple flat-crowned, wide-brimmed *sombrero* (hat) made of leather, palm fiber, or felt. Many rode barefoot, but others wore leather shoes, with or without heels. A bandanna and iron spurs with large rowels completed the apparel.[13]

Vaqueros developed their principal tools in the sixteenth and seventeenth centuries. The heavy Spanish *silla de montar* (military saddle) and light Moorish *jineta* evolved into the stock saddle. The *desjarretadera,* or hocking knife, was a crescent-shaped blade sharpened on the inside of the curve and mounted on a pole ten or twelve feet long. A rider following a cow or bull could thrust the blade against the back of the leg, cutting the hamstring and bringing the animal down. Another thrust at the back of the neck severed the spinal cord, and the animal could be skinned for its hide and tallow. Though the desjarretadera was outlawed in 1574 because its use was leading to a depletion of the herds, vaqueros along the northern frontier continued to use it for two centuries more. The other important tool of the vaquero was the *lazo* (lasso) or the *reata* (lariat), a braided rawhide rope with a large loop at the end. The vaquero initially placed the loop of the lasso over the end of a long lance, the *garrocha,* then used the lance to guide the loop over an animal's head. This method was eventually displaced when vaqueros learned to throw the lasso, a technique that could be used from the back of a running horse or on foot in a corral.[14]

Horses were indispensable to the livestock industry. Those brought from Spain were of mixed quality, not the Andalusian bloodlines prized throughout Europe in the sixteenth century. As the Indians of the north learned the value of the horse, they found ways to breed and trade them. By 1600 horses were being raised in what is now New Mexico, and by the middle of the seventeenth century, the Apaches of the southern plains were using them. By 1673 they were reported among the Indians near the confluence of the Missouri and Mississippi rivers. The Caddos living in what would become northeastern Texas and adjacent Arkansas, Louisiana, and Oklahoma, acquired horses by the 1680s. The Comanches were raising them on the Great Plains by about 1700. Horses became abundant in Texas in the eighteenth century.[15]

Frontier missions of the Catholic Church brought farming and ranching to northern New Spain in the seventeenth century. The missions were managed solely for the benefit of the Indian neophytes, and life was simple but well organized. The missionaries, and sometimes soldiers, taught the Indians the rudiments of Christianity and the skills needed to plant and care for crops, herd livestock, and manufacture many of the simple goods and tools they needed. Missions were designed to be temporary institutions. Within a generation of being established, they were supposed to attract a stable population of Indians, instruct them in the Catholic faith, and transfer mission activities and responsibilities to parish clergy and the secular town. The objective was to produce a stable community of parishioners and resident landowners, or *vecinos.* Upon completion of their task, which often took longer than a single generation, the missionaries moved to a new frontier.

Throughout New Spain, the army was charged with helping the clerics establish, maintain, and protect the missions from hostile Indians.

Presidios were constructed far enough from the missions to avoid conflict with the Indian residents, but close enough to provide military protection in case of need. Two or three soldiers usually lived at each mission to help train and supervise the Indian residents. The other presidial soldiers kept busy patrolling the area, guarding herds of livestock, escorting the mail and travelers from one settlement to another, and assisting in construction projects. Soon after the presidio and one or more missions were established, a town, or *villa,* usually sprang up to supply their needs and benefit from the protection they provided. The initial residents were often traders, artisans, retired soldiers, and ranchers.

FRENCH AND SPANISH EXPEDITIONS INTO TEXAS

By the 1680s the northeastern frontier of New Spain reached from El Paso del Norte on the Río Grande; to Santa Barbara and Parral on the Río Conchos; to Saltillo; to Monterrey at the foot of the Sierra Madre Oriental; and to Tampico on the Gulf of Mexico. The Pueblo rebellion of 1680 in Santa Fe had thrown the Spaniards out of New Mexico, and they appeared to have little energy left to push farther north into what is now Texas. But this would change dramatically in 1685 when René-Robert Cavelier, Sieur de La Salle, attempted to establish a French colony near the mouth of the Mississippi. Hoping to strengthen French claims to the Mississippi and from there threaten Spanish mines in Mexico, La Salle led a well-supplied expedition. However, his three ships completely missed the Mississippi and entered Matagorda Bay. Immediately beset by misfortune, including shipwrecks, lost supplies, rebellion, disease, and hostile Karankawa Indians, the remnants of the expedition were able to construct a crude wooden palisade and five houses, which they named Fort St. Louis, on Garcitas Creek near present-day Victoria, Texas. In desperation, La Salle set out with a small party to find the Mississippi River, only to be killed by his own men before leaving Texas. Though the French tried to maintain strict secrecy about the La Salle expedition, rumors soon reached New Spain. To counter this threat, beginning in 1686 Spanish authorities organized a series of expeditions, or *entradas,* into Texas. Led by Alonzo de León, the Spaniards finally found the remains of Fort St. Louis in 1689, where only some of the children had survived a recent attack by the Karankawas.[16]

Still wary of French intentions, de León's next expedition, in 1690, established Mission San Francisco de los Tejas among the powerful

Caddos of eastern Texas and western Louisiana. However, the well-fed Caddos saw no reason to congregate at the missions. The Spaniards were unable to build a functional irrigation system because of the hilly terrain, so crops failed and food shortages occurred. The Indians stole cattle and resented the soldiers' interest in the Caddo women. Epidemics killed both the Indians and the Spaniards. Despite efforts to resupply and strengthen the mission, by the autumn of 1693 the situation was hopeless, and San Francisco de los Tejas was abandoned. But Spain remained interested in the northern frontier. Between 1700 and 1703, in an attempt to convert the Indians along the Río Grande, the Spaniards established three missions and Presidio San Juan Bautista del Río Grande about thirty-five miles south of present-day Eagle Pass.[17]

In 1699 the French finally located the mouth of the Mississippi and established the post of Biloxi. In 1700 Louis Juchereau de Saint-Denis, a Canada-born, Paris-educated adventurer, arrived there. In 1713 he established a trading post in Natchitoches. Initially trading with the Caddos for livestock and buffalo hides, Saint-Denis concluded that trade would be much more profitable if the Spaniards reestablished their missions in eastern Texas. In 1715 he set out to visit San Juan Bautista, the nearest Spanish settlement. Crossing the Texas coastal plains, he reported that Spanish livestock, which had strayed from earlier entradas or had been taken by the Indians, "count by now the thousands, cattle as well as horses, and the land is overrun with them." Soon after arriving at San Juan Bautista with French trade goods, Saint-Denis married into the influential Spanish family of Commandant Diego Ramón. In the spring of 1716 Saint-Denis and Ramón's son, Captain Domingo Ramón, led an entrada of twelve priests, twenty-five mounted soldiers, and forty civilians, including several women, into the lands of the Caddos. Despite the abundance of wild livestock, the entrada took along an additional sixty-four oxen, five hundred horses, and more than one thousand sheep and goats. There, in 1716 and 1717, the Spaniards established six small missions and a presidio in hopes of finally attracting and converting the powerful Caddos.[18]

THE CADDOS

The southwesternmost representatives of the great Mississippian cultures of middle North America, the Caddos were known as the *Tejas* to Spanish missionaries. For almost a millennium, the Caddos dominated their homelands of northeastern Texas and adjacent areas of Arkansas,

Louisiana, and Oklahoma. By the late 1600s European diseases, transmitted via extensive Indian trade routes, had greatly reduced Caddo populations, probably to less than 40,000. They were organized into two chiefdoms, the Kadohadachos along the Red River north of present Texarkana, and the Hasinai who lived to the southwest along the Sabine, Angelina, and Neches rivers. Each chiefdom was ruled by a priest-chief, the xinesí, and was comprised of several city-states, each controlled by a Caddí who answered to the xinesí. All the Caddos spoke Caddoan, a language related to those spoken by the Wichitas and Pawnees to the northwest. The Caddo economy was based on a productive, unirrigated, slash-and-burn farming system supplemented by extensive hunting, fishing, and gathering of wild plant materials. Their prinipal crops were the "three sisters," corn, beans, and squash; and they lived in semipermanent villages near permanent streams and their gardens and fields. A village with 150 inhabitants might have had eight to twelve large houses located at intervals of a few hundred yards along stream terraces or uplands near a permanent watercourse "where there is water at hand for household use and for bathing—which is very frequent among them all." Surrounded by a few outbuildings, middens, and outdoor work areas, the houses would have been only a short walk from the fields where corn, beans, pumpkins, squash, and watermelons were grown.[19]

Throughout the southeastern United States, American Indians cleared forestlands for cornfields. The land clearing and agricultural practices were probably similar throughout the region. First, underbrush was burned, and trees were killed by girdling the bark with a hatchet. As the tree died, dead limbs were piled and burned at its base to hasten its fall. Except in drought years, the Caddos produced abundant corn, beans, squash or pumpkins, melons, sunflowers, and tobacco. Forest openings and prairies, especially those on sandy and sandy loam soils in the river bottoms, were the basis of Caddo agriculture. They planted crops in these openings for several years until soil fertility declined or weeds became too difficult to control. The Caddo farmers then moved on to other openings nearby, leaving the exhausted fields to rest and regain their fertility.[20]

The planting season began in March when fields were cleared by burning. Tillage began immediately. It involved a team of men and women who worked together "breaking up just the surface of the earth with a sort of wooden instrument, like a little pickaxe, which they make by splitting the end of a thick piece of wood, that serves for the handle, and putting another piece of wood, sharp pointed at one end, into the slit." After finishing their work, usually in two or three hours, "the owners of the house give them an abundance of food. They then move to another spot to do the same thing."[21]

Indians throughout what is now the southeastern United States planted corn in rows three or four feet apart with a similar distance between groups of plants in the row. For each group, the farmer made a hole with a long pointed stick or a hoe and planted three to five seeds. The farmer then covered the seeds and left them to germinate and emerge. When the corn was about a foot tall, women and children weeded the crop in the first of three or four passes. During the weeding process they pulled soil into hills around the plants to provide support.[22]

The Caddos, like other southeastern tribes, planted two kinds of corn. The first, the "small corn," was a short-season, popcornlike variety, "the stalk of which is not more than a *vara* [thirty-three inches] in height. However, it is covered from top to bottom with ears which are very small but covered with grain." Planted after the threat of frost had passed, the crop was weeded once while the plants were small. This early-maturing variety normally escaped the mid- and late-summer droughts and was probably eaten both fresh and dry. "Upon the same ground, after clearing it anew," they planted a larger, longer-season flour corn, which was harvested in the late summer and fall.[23]

Most southeastern Indians planted beans and squash between the corn plants after the corn had emerged. The beans were often allowed to use the corn stalks for support. The Caddos also planted their beans "in an odd way." "In order that the vines may run and be protected from small animals and from mildew, they stick a forked cane at each hill. Thus the vine bears more abundantly and it is no trouble for them to gather the crop because they pull up the cane and carry the whole thing home." In addition, their sunflowers grew to be "quite large." The flower was "enormous" with seed "like the piñon."[24]

The Caddos gathered a wide array of wild foods to complement their crops. Berries, plums, and grapes ripened in the summer, and acorns and nuts were abundant in the autumn. Fray Isidro Felis de Espinosa, who participated in several entradas during the early 1700s, noted that "the entire country is filled with various kinds of trees, such as oaks, pines, cottonwoods, live-oaks, large nuts—which yield the thick shelled nuts—and another kind of tree which yields small thin shelled nuts. The Indians use all these as food." Archaeological sites in eastern Texas contain the remains of corn, squash, acorns, hickory nuts, pecans, black walnuts, and seeds of grapes, sumac, honey locust, water locust, and hackberry.[25]

The timberlands of eastern Texas were home to an abundance of wildlife, including deer, turkeys, panthers, and black bears. The river bottoms were excellent habitat for raccoon, opossums, mink, gray foxes, beavers, black bears, gray squirrels, and turkeys. Buffalo were hunted on the prairies to the west. Each autumn, the skies filled with waterfowl, hawks, and other migratory birds as they made their way southward to the coast or beyond. Caddo men were expert hunters and killed a wide variety of game. Archaeological excavations have revealed that between the years 800 and 1300 the Caddos made particularly heavy use of deer, turkeys, box turtles, and to a lesser extent, cottontail, swamp, and jackrabbits. Espinosa reported, "There are lagoons in which an abundance of fish are found. . . . When warm weather comes the Indians go with their families to certain spots and stay for some days, living on fish. They carry home quantities of cooked fish, I ate some of these."[26]

The buffalo hunt normally occurred in November and December and involved most of the population, who moved to the hunting camps and lived in skin tents like the ones used by Plains Indians. In the 1720s Espinosa reported, "The buffalo is distant more than forty leagues [104 miles] from the [Caddo] country, and to secure a supply of dried meat the Indians all go well armed because at this time if they fall in with the Apaches the two murder each other unmercifully."[27]

Prior to their acquisition of European firearms, the Caddos were renowned for their use of the bow. The bow and arrow had replaced the atlatl, or spear thrower, by about the year 800. In 1806 Peter Custis reported that the Caddos' "principal weapon is the Bow & Arrow which they wield with astonishing dexterity & force.—It is said they can with great ease throw the Arrow entirely through a Buffalo. It would be worth a journey of two or three thousand miles to see them use it."[28]

The Caddos, like the Plains Indians, adopted horses as soon as they became available, using them for transportation, hunting, and war. By the late 1680s horses were plentiful among the Caddos, and La Salle traded a metal ax for one. By the 1720s Espinosa could report that the Tejas "ride on horseback with great skill, their feet hanging loose and, traveling at a great rate, they guide their horses with only a slender cord which they use in place of a bridle."[29]

During the late seventeenth and eighteenth centuries the Caddos traded large amounts of furs and hides to the French and Spaniards. Most of these were exported through Louisiana. For example, the Spanish general Terán reported that in less than a year during the early 1690s, the Caddos brought the skins of forty thousand deer, fifteen hundred bears, twelve hundred otters, and six hundred beavers to Nacogdoches for sale to the French. Most years, the Caddos went north during the winter, where they killed "a great number of bears" and brought back "a great deal of bear fat rolled up in moss and loaded on their horses." After rendering the fat, they kept it "in pots for seasoning for the whole year. . . . It is certainly true that they need nothing else for seasoning when they are supplied with this." As a result of this trade with the Caddos and other tribes, the value of pelts exported from Louisiana increased from about 3,000 livres in 1718 to 120,000 livres in 1756, and to 250,000 livres in 1762.[30]

Caddo women were responsible for all aspects of food production and preparation except hunting and preparing the soil. By about 1450 corn was the most important staple of the Caddo diet. The grain was carefully stored to assure an adequate supply for both food and seed corn. To protect their seed corn, the Caddos made "a string of the best ears of grain, leaving the shucks on, and put it up on a forked stick at a point in the house" where the smoke would deter insects. Enough seed corn was stored to provide for two years' plantings "so that, if the first year is dry, they will not lack seed for the second year." To prevent weevil damage, the Caddos stored their corn and beans in "large baskets made of heavy reeds" and then covered the grain with "a thick layer of ashes."[31]

Caddos used both fresh and dried corn in a variety of dishes. The women scraped fresh corn from the cob, probably using mussel shells from the local streams. The resulting mush could be fried or boiled and consumed immediately. Sometimes, grease was added to the mush, and the mixture was baked. The resulting cake was then broken into pieces, dried on top of an arbor or roof for several days, and stored in sacks hung inside the storage building. A mush was prepared by boiling one part cake in three parts water, with or without dried or fresh meat. Like other American Indian farmers, the Caddos ground dried corn with mortars made from the trunks of trees. These were buried upright in the ground, and the upper ends were "excavated by means of fire to a certain depth." As many as four women would beat the corn with wooden pestles "about five feet long, and they preserve[d] a cadence in the way the blacksmiths beat on their anvils." After pounding the corn "for a certain time, they [took] out the said meal and other women pass[ed] it through little sieves which they [made] very neatly out of large canes." When finer meal was needed, they used "little winnowing baskets . . . in which the finest remains caught on the bottom; the grits and the bran come out above." Cornmeal was prepared in a number of ways: it was eaten alone as a fine dry powder, mixed with beans and boiled, used to thicken stews and soups, mixed with honey, or formed into a patty and boiled with or without dried meat. Caddo women also prepared a thick paste of ground roasted corn and sunflower seeds called *bajan.* The French traveler Pierre Pagés wrote that the best provision for traveling was "a piece of flesh dried in the sun, and a small quantity of ground Indian corn, named by the Spaniards *pynolé* [pinole]. This meal, when mixed with water, swells to a great bulk, insomuch that a single handful of it suffices amply for one repast."[32]

The Caddos planted "five or six kinds of beans—all of them very good, also calabashes, watermelons, and sunflowers. The seed of all these, mixed with corn make very fine *tamales.* They also use another kind of seed like cabbage seed which, ground with corn, make[s] a kind of meal." They prepared beans by "placing them in a big pot without removing the strings even, since they cover them with vine leaves, until they are almost cooked." They served the beans "in a bark apron" and shelled them as they ate. The women boiled dried pieces of pumpkin with meat and sometimes added ground or boiled nuts to the mixture. They dried and stored wild plums and grapes for future use. Sunflower seeds were also ground, mixed with cornmeal, and made into small cakes or tamales.[33]

Caddo women placed hickory and pecan nuts in the shallow depression of a nutting stone and smashed them with a stone *mano.* They then boiled the mixture and skimmed the oil off the top for seasoning. They boiled whole acorns in water, mashed them, and leached the paste with water until the tannic acid had been removed. They mixed the remaining white paste, as well as the ground kernels of pecans and other nuts, with corn flour to make breads and puddings. Both the nuts and acorns contained protein, and the acorns were a good source of lysine, a valuable amino acid that is deficient in corn. Archaeologists have also found carbonized seeds of lamb's-quarters, hackberry, and smartweed at Caddo sites. The Caddo women probably parched, ground, and mixed these seeds of wild plants with other grains and meat. The mixture was then used to prepare mush or stew.[34]

Fresh meat was also cooked or dried for later use. The Caddos living in Oklahoma in the 1930s dried meat the traditional way, by cutting it into thin slabs "just as thin as it could be cut." "This would then be hung on a rope, or sticks, and turned over once or twice a day. It usually took about three or four days to dry, depending on the weather. The meat would keep indefinitely when dried, and was good to eat raw." In addition, the Caddos often "roasted beef on coals of fire, then put it in a mortar and pounded it up with a pestle" before boiling it in a pot with a little water. They sometimes had two kinds of meat at a meal, "one boiled and

Caddo house

the other roasted," which were served on "very pretty platters" made of reeds. Meals were social occasions, and Fray Francisco Casañas, who helped establish the first missions in East Texas, observed that "they take a long time to eat and while they are eating, they sing and talk, and, from time to time, whistle. . . . After eating, the guests are supplied with the requisites of smoking."[35]

The Caddos produced a wide variety of pottery and basketry products, most of which were used in food preparation. Throughout their history, Caddo women made both plain and finely decorated ceramics. Distinctive patterns and styles distinguished specific communities and craftswomen. Many types of coarse and finely woven baskets and mats were made, each for a special use. Raw materials included "willow, hackberry, slippery elm, dogwood, cattail, swamp grall, soap weed, or bear grass. Hackberry was used mostly for sifting corn and washing lye hominy . . . because it has no bad taste."[36]

Because of the stability provided by their agriculture, the Caddos could afford to construct well-built, dome-shaped houses made of wooden frames covered with grass. In 1687 Henri Joutel described them as "round at the top, after the manner of a bee-hive or a rick of hay. Some of them are sixty feet in diameter." He noted that several families could live in a single house, and "in some of them there are fifteen or twenty [people], each [family] of which has its nook or corner, bed and other utensils to itself, but without any partition to separate it from the rest. However, they have nothing in common beside[s] the fire, which is in the midst of the hut, and never goes out." A wooden mortar for grinding corn was placed between the fire pit and the western door.[37]

For hundreds of years the Caddos were the most highly organized and successful people in what is now Texas. Taking full advantage of the natural resources of their homeland, they developed an economy based on hunting the region's abundant wildlife and gathering wild foods. The Caddo food system was reliable, sustainable, and its communal nature gave the Caddos and other Mississippian peoples the leisure to develop complex hierarchical societies with impressive material cultures. If the introduction of European diseases had not begun to decimate Caddo populations, the confederacies would have probably successfully incorporated European technologies and would have remained a powerful political force well into the nineteenth century. However, as Caddo populations were decreasing under the pressure of epidemic diseases, another type of agriculture and rural economy was pushing northward from what is now central Mexico.

By the early eighteenth century New Spain, in response to French threats from Louisiana, had established a toehold in eastern Texas among the powerful Caddo nation. Entirely dependent on their powerful and well-fed hosts for food, the Spaniards had not begun to develop a sustainable agriculture in Texas. But the distance between the eastern Texas missions and San Juan Bautista on the Río Grande was too great, and Spain established a way station between the two. This station, initially consisting of Presidio San Antonio de Béxar and Mission Valero, would grow into an agricultural and ranching community of five missions, a villa, and the presidio. Known simply as "Béxar" or "San Antonio," this settlement would play a large role in Spain's, and later Mexico's, fortunes in Texas.

Chapter 2

MISSIONS AND FARMS ON THE RÍO SAN ANTONIO

In 1716 and 1717 the new entrada reestablished six small missions in eastern Texas. But the Caddoan tribes, stronger militarily and far better fed and housed than the Spaniards, still saw no reason to congregate at the missions. The region's deeply dissected rolling lands could not be irrigated. In addition, the lack of Caddoan laborers at the missions made large-scale dryland farming and major construction projects impossible. Because of deteriorating conditions at the new missions, in 1717 and 1718 entradas were organized to resupply them. It soon became clear that a way station was needed to break up the long journey between San Juan Bautista and the Tejas missions. As a result, in the spring of 1718, Father Antonio Olivares and other Franciscans from the College of Santa Cruz de Querétaro on their way to eastern Texas paused near the headspring of the Río San Antonio. There they established Mission San Antonio de Valero, later known as the Alamo. Temporary structures of brush, mud, and straw were constructed to house the first inhabitants of the mission, several Indians that Father Olivares had raised from childhood. Hispanicized Coahuiltecan Indians were also brought from northern Mexico because their language and culture were similar to those of Texas Coahuiltecans, hunters and gatherers without the strong social organization, rich material culture, or military prowess of the Caddos. Four days later Presidio de San Antonio de Béxar was established about one mile to the north by the military leader of the expedition, Martín de Alarcón. A few families accompanying the entrada settled around the presidio, establishing a small civilian settlement.[1]

MISSION LANDS AND HERDS

But why did Olivares and Alarcón select the valley of the Río San Antonio as the site of their way station between San Juan Bautista and east Texas? Clearly, the Río San Antonio was attractive because it was the first river north of the Río Grande with an abundant and permanent supply of irrigation water. San Pedro Springs fed San Pedro Creek, whose crystal clear water ran through the valley "with a current of not less than five or six miles an hour, over a bed of rocks." Some two miles to the northeast, the headspring of the Río San Antonio provided even more water than San Pedro Springs. Near Mission Valero, the river was described as "thirty yards in width and from three to four feet in average depth." After flowing southward for about five miles, past Mission Valero and the presidio, San Pedro Creek and the Río San Antonio came together. The missionaries realized that the slope of the river valley would allow them to irrigate "by throwing a dam across the bed of the stream [San Pedro Creek], to carry the water by a ditch along the upper part of the mission farm and, by such means, suffer it to run over every portion of it as upon an inclined plane."[2]

The 1718 expedition to resupply the east Texas missions included seventy-two persons, 548 horses, six droves of mules, and other livestock. Despite these efforts, conditions deteriorated in 1718–19, and the missions were again abandoned. The retreating inhabitants, arriving in San Antonio in the autumn of 1719, realized that the valley's warm climate, abundant irrigation water, and fertile soil made it an ideal place to settle, and in 1720 Mission San José y San Miguel de Aguayo was established on the west side of the river about three leagues below Valero (a league equals about 2.6 miles). The mission was named for the Marqués de Aguayo, the wealthy Spanish governor and captain general of Coahuila and Texas, who had recently obtained permission to lead an even larger entrada in a third attempt to establish the east Texas missions and strengthen Spain's presence in the region. After sending troops ahead to Matagorda Bay and San Antonio, Aguayo arrived in San Antonio in March 1721, bringing with him five hundred soldiers, twenty-eight horses, forty-eight hundred cattle, and sixty-four hundred sheep and goats. Moving quickly to east Texas during the summer and autumn of 1721, Aguayo accepted the truce then in effect between France and Spain. He reestablished Presidios Nuestra Señora del Pilar de los Adaes and Nuestra Señora de los Dolores (commonly called Presidio de los Tejas), as well as the six missions abandoned two years earlier. In the spring of 1722, determined to thwart any remaining French designs on Texas, Aguayo moved southwest to Matagorda Bay, where he founded Presidio Nuestra Señora de Loreto, precisely on the site of La Salle's Fort St. Louis. Nearby he established mission Espíritu Santo de Zuñiga to congregate Karankawan and Coahuiltecan groups. The mission and presidio would become widely known as La Bahía, or "the Bay." Aguayo's entrada greatly strengthened Spain's influence in Texas, leaving three areas of substantial Spanish settlement: east Texas, Matagorda Bay, and the valley of the Río San Antonio, with a total military presence of more than 250 soldiers.[3]

At Valero and San José the missionaries immediately began to attract Coahuiltecan Indians with promises of safety and a steady supply of food. With the help of interpreters, missionaries worked individually with new arrivals to teach them the fundamentals of Christianity and train them to work within the missions. It was clear from the beginning that the fruit, vegetable, and cotton crops needed by the missions, presidio, and civil settlement could not be reliably produced without irrigation. The missionaries used the knowledge gained from Spain's long history of irrigated agriculture to begin constructing the community's first irrigation ditch, or *acequia,* to serve the Valero mission. The acequia, later called the Alamo *madre* ditch, was begun in 1718 and irrigated mission lands along the east side of the river. It began at the headsprings of the Río San Antonio, ran about three miles along the east side of the river to Mission Valero, then rejoined the river below the presidio. In the early 1720s the presidio was moved to a more secure location east of San Pedro Creek and west of the horseshoe bend in the river. During this period the soldiers, assisted by their *vecinos* (civilian neighbors) and mission Indians, began construction of the San Pedro acequia, also known as the *acequia madre.* This canal irrigated about four hundred acres of "fertile land found within the angle formed by the San Pedro [Creek] and the San Antonio [River], taking water from the former for the benefit of the presidial troops and settlers that might join them." That is, it took water from San Pedro Springs almost two miles north of the presidio, carried it along the divide between the creek and the river, passed through the civilian settlement just east of the presidio, and continued southward for about two more miles until it emptied into the river. The San José acequia diverted water from the river a short distance below the mouth of

San Pedro Creek, conducted it to the west of the river past the mission, and irrigated about six hundred acres of land west of the river. A map drawn in the late 1720s shows acequias watering corn and wheat fields near the presidio and Missions Valero and San José (see Map 2).[4]

By the mid-1720s peace in Europe had reduced the threat of French incursions. As a result, in 1727 and 1728 Brigadier General Pedro de Rivera y Villalón inspected the Texas presidios, seeking ways to reduce the costs of maintaining the empire. He recommended that Presidio de los Tejas in east Texas be closed and that the number of soldiers at the other presidios be reduced. Protesting that the missions' safety was threatened, in the summer of 1730 three east Texas missions moved to what is now known as Barton Springs in present-day Austin. One year later the missions—Nuestra Señora de la Purísima Concepción, San Juan de Capistrano, and San Francisco de la Espada—relocated to the valley of the San Antonio. Thus, by the 1730s, five missions were ranged along the Río San Antonio about one to three miles apart. They were, from northwest to southeast, Valero, Concepción, San José, San Juan, and Espada. In 1731 these five missions were awarded large grants of land to pasture their herds. Valero's pasturelands, like the mission itself, were on the east side of the river, extending from the east around to the northwest of the mission. A short distance down the river on the same side was Concepción, whose ranch extended eastward to the Río Cíbolo and covered about fifteen leagues. San José, farther downriver on the west side, had its ranch to the west and southwest. San Juan Capistrano, on the east side of the river, had lands to the south of Concepción's, extending to the Río Cíbolo on the east. Espada, farther down on the west side of the river, held lands south of San José's ranch (see Map 2).[5]

Realizing that irrigation was necessary for the missions' survival, the Crown awarded water rights to each of the missions. Because they were located on both sides of the river a mile or more apart, each mission developed its own acequia with all the necessary dams, lateral canals, and other structures needed to irrigate its fields. Mission Concepción's acequia was built in 1729. It took its water from the horseshoe bend of the Río San Antonio and watered lands east of the river below the presidio and civilian settlement. Because the dam was low, only about five feet tall, a cut of about fifteen feet was needed to draw water from the river. But after the river traveled a considerable distance below the level of the surrounding fields, the soil surface fell to the level of the water, which could then be used for irrigation. Tradition holds that boats were used on this, the largest of the acequias, for transportation between the civilian settlement and Mission Concepción. The San Juan acequia was begun in

MAP 2. *San Antonio missions and acequias.*

1731. It took water from the river opposite Mission San José and irrigated about five hundred acres on the east side of the river. The Espada acequia, which must have been begun soon after the mission was established in 1730, was three miles long and five feet wide and irrigated about four hundred acres on the west side of the river (see Map 2).[6]

Mission Indians, fed by the clergy and closely supervised by soldiers, provided most of the labor needed to dig the mission acequias. To set the grade of the ditch, missionaries probably used levels consisting of an inverted equilateral triangle with a plumb line suspended from the middle of the upper side. When the triangle was adjusted so that the plumb line intersected the downward pointed apex, the upper side of the triangle provided a level line of sight. Mission Indians dug the acequias by hand with bars, picks, and shovels, which were sometimes fashioned of wood. They carried earth and rocks on hides dragged by oxen or suspended from poles between two workers. The acequias normally wound around obstacles, such as mounds of earth and large trees, so that workers could avoid having to move large amounts of soil in constructing cuts and fills. Though most of the ditches were probably unlined, the vertical walls of the Valero acequia south of the mission were faced with quarried limestone. One six-foot- wide, five-foot-deep section was lined with five layers of cut stones, each ten to fourteen inches thick and eleven to forty-one inches long. Dams used to divert water into the canals were usually constructed of loose rock, brush, and mud, and they required constant repair. Only a few, such as those at Espada, were made of stone. Canals usually crossed small gullies or other canals by means of hollowed-out logs, or *canoas,* through which water flowed. One exception was the arched stone aqueduct that conducted the Espada acequia across an arroyo. Small intermittent streams sometimes dumped storm water directly into the acequia. To prevent these large flows from overtopping and destroying the earthen walls of the canals, workers constructed masonry sections to allow excess water to flow safely over the side. Headgates used to divert water from the canal into the fields were made of wood. The mission Indians cleaned the acequias in the spring by scooping out accumulated silt with paddlelike shovels made of walnut, a wood that allowed the mud to slip cleanly off the blade when it was dumped on the ditch banks.[7]

CONFLICTS WITH THE APACHES

When the Spaniards founded San Antonio in 1718, the eastern Apaches lived along the upper reaches of the Río Brazos and Río Colorado. The Spanish horses proved an irresistible temptation, and in 1720 the Apaches attacked supply trains, killing soldiers, civilians, and mission Indians to obtain the horses. Three years later eighty horses were stolen from the presidio's herd, but an expedition of sixty soldiers and mission Indians pursued the Apaches for over three hundred miles, capturing hostages and the Indians' horse herd. Though the late 1720s were a time of relative peace, hostilities soon increased. In 1730 a large band of Apaches attacked the presidio, killing two soldiers and stealing sixty cattle. The Spaniards responded with a punitive raid on an Indian camp west of San Antonio, killing a large number of women, children, and warriors. In September 1731 a major battle involving about 20 Spaniards and 500 Apaches and their allies occurred less than two leagues from the presidio. In a pitched battle 2 Spaniards were killed, and 13 were badly wounded. In December 1732, a force of 157 Spaniards and 60 mission Indians set off with 140 pack animals and 900 horses and mules. On the Río San Saba they surprised four Apache villages, or *rancherias,* consisting of about 400 tents and 700 warriors. Attacking at dawn, the Spanish guns killed an estimated 200 Indians in a five-hour battle. They also took about 30 captives, 700 horses, and 100 loads of hides and other property.[8]

Troubles continued, and by the late 1730s, the vecinos and soldiers were forced to bring their horses into the city every night to protect them from the Apaches. José de Urrutia, governor and captain of the presidio, wrote that "those [Apaches] who can enter a presidio at night as far as the center of the plaza and who without being heard can safely remove the horses from the corral in which they are tied to the doors of the houses are to be feared." Even during the day, vigilance was necessary. Herders seldom went "beyond the edge of town without a gun" or ventured "more than a mile from town on account of the Indians who often lurk in the vicinity watching an opportunity to cut off their retreat and drive off the horses under their charge." Because it was dangerous for the mission Indians to herd their livestock, more and more cattle grew wild. By the mid-1730s *manso,* or tame, cattle were scarce, though unbranded cattle, or *mesteños,* were plentiful. Noting that the population of the villa needed meat, the governor ordered that the alcalde organize parties of six hunters, or *carneadores,* with four soldiers for protection to provide beef for the town.[9]

IRRIGATED FIELDS FOR THE VILLA

In an attempt to increase the vigor of the civilian settlement, sixteen families from the Canary Islands, called *isleños,* were brought to San Antonio in early March 1731. Despite the fact that they had had no such standing in their native land, upon reaching San Antonio they were declared persons of noble lineage, or *hidalgos.* As a result of the islanders' arrival, the civilian settlement near the presidio was raised to the status of a villa, San Fernando de Béxar, located between San Pedro Creek and the river, about the distance of a musket shot west of the presidio. Because the irrigated land between the creek and the river south of the presidio had been farmed communally, no one held formal title to it. Each of the sixteen isleño families was granted a parcel of this farmland, and fields were quickly cleared and plowed. Within a few weeks, the isleños had planted twenty-two bushels of corn, as well as beans, oats, cotton, melons, chilies, watermelons, pumpkins, and other vegetables.[10]

The land was divided into long, narrow fields reaching from San Pedro Creek to the river, each of which could be watered in a single day using the entire flow of the San Pedro acequia. Because a lottery was used to determine which family got a particular parcel of land, the term *suerte,* referring to the luck of the draw, came to mean a parcel of irrigated land. The sixteen suertes initially awarded to the isleño families in 1731 were 105 varas wide (a vara equals approximately 33 inches), and the lengths varied according to the distance from the acequia to the river or creek. In 1734 additional farmland south of the presidio was granted so that each isleño family had two suertes. The irrigated land below the presidio thus became known as the *labor de los isleños* or the *labor de abajo.* Because the soldiers and vecinos who had cleared the fields, dug the acequia, and farmed the land prior to the arrival of the isleños held no formal titles, they were left with neither land nor irrigation rights.[11]

The water of the San Pedro acequia was divided into twenty irrigation days, or *dulas,* to ensure equitable distribution of water among the fields. Each of the sixteen isleño families received one dula (twenty-four hours) of water, and the town council received four, which it could lease to raise funds. Water rights, like land, could be bought, sold, and inherited. As a result, the original suertes and dulas soon began to change hands, and it became common for individuals to own fractions of both, such as one-half suerte of land and one-fourth dula (six hours) of irrigation water. Owners of suertes were required to keep the ditch and its gates clean and in good repair, and penalties could be imposed for fouling or obstructing the ditch, stealing water, or taking it out of turn. The mayordomo, or water master, was authorized by the town to police the use of water from the San Pedro acequia, prevent its waste, and inspect irrigation gates. Herds were not allowed to cross the irrigated fields to drink from the acequia madre.[12]

The Béxar Archives record the sale of several suertes of irrigated farmland from the late 1730s to the late 1740s. In many cases two suertes of irrigated land belonging to an individual were sold together. Prices for the two varied from 200 to 333 pesos, with payment usually in livestock and effects rather than cash. In contrast, prices for lots with stone houses ranged from 120 pesos to over 800 pesos. Lots without houses usually sold for less than 100 pesos.[13]

CONFLICTS

One of the most important privileges granted to the isleños was full control of the villa's *cabildo,* or city council. They initially used this political advantage to maintain control of virtually all nonmission lands west of the Río San Antonio, as well as most of the water rights of the San Pedro acequia. The cabildo consisted of four *regidores* (councilors) and two *alcaldes ordinarios* (magistrates). The seats were originally appointed by the king through his governors but were later passed on by purchase, by inheritance, or as gifts. As in other colonial towns, a few influential families controlled the government of San Fernando. This abuse of power created divisions between the isleños and other civilian settlers that lasted for decades.[14]

In the 1730s members of the cabildo and other prominent isleños were involved in a number of civil suits and edicts related to crop and livestock production. In June 1733 Joseph Padrón alleged that councilman Juan Leal had ordered two hired laborers to plow up a small part of

Padrón's immature corn crop. Leal responded that because the boundaries between the suertes had not been correctly laid out, Padrón had actually planted on Leal's land, and the only cornstalks that had been destroyed were on Leal's land at the end of the rows where the oxen had turned. After substantial testimony the alcalde ruled that Leal had previously cut an irrigation ditch on Padrón's land and had used land belonging to several other isleños. Leal was placed under house arrest and was eventually ordered to pay a fine of twenty-five pesos or to have some of his cattle sold "at auction in the public square." The lawsuit dragged on until March 1734 when the parties decided that because of the "evil consequences resulting from lawsuits and disputes," they would agree to "absolute peace" and "live in unity and harmony."[15]

Most families living in the presidio and villa kept a few cattle, horses, oxen, or mules. These were often let out of their pens to graze, and they were naturally attracted to the irrigated crops of the isleños. The irrigated land below the presidio was protected on the east, west, and south by San Pedro Creek and the Río San Antonio. However, livestock could enter from the north and soon began to cause considerable damage. As a result, in 1732 a fence was built along the north side of the irrigated land. But by 1735 it was in disrepair, and cattle and horses were entering the fields to graze. As a result, in April 1735 Leal, now the alcalde, ordered the isleños to repair the fence within fifteen days. In addition, he ordered "all residents and stock breeders who have any horses or cattle to place them in charge of a herder who shall take care of them during the day and shut them up at night." Anyone whose livestock got into the irrigated fields would be fined. But the sixteen isleño families immediately lodged a protest with the governor, saying that they did not want to build the fence "in the place determined by the . . . alcalde because it would make things too inconvenient for them." After a vote "on whether a fence should be built and where it should be placed" resulted in a tie, with eight votes for each of two possible sites, the governor ordered the villa residents to build a fence 145 *brasas* long (one brasa equals two varas, about 66 inches) along the edge of the villa within five days.[16]

Strengthening the fence on the north side of the irrigated fields was apparently insufficient to keep livestock out, and in late September 1737, the governor reported that the farmers were suffering "serious losses . . . because of the continued damage done to their cornfields by the unherded cattle of the missions and of the soldiers and residents of this royal presidio." In addition, the missionaries, mission Indians, and soldiers complained that the farmers living in the villa had "killed some of their cattle and beaten up others for damaging their cornfields, which were unprotected by fences." The governor responded by ordering once again that all cattle "be placed in charge of herders who will keep them away from the cornfields." The farmers were again commanded to "fence their cornfields as required by royal orders." The governor reminded all his charges that they were living in "a land where the enemy is always in sight." Only by "being united in will" could they "repel the enemy, thwart their continued assaults and at the same time make progress with their crops," which "because of the remoteness of this land . . . is the only possible source . . . of food for all." Members of the cabildo, representing the farmers of the villa, immediately protested that they had tried to correct the situation by "taking the cattle to their owners, by keeping them penned in the corrals for three or four days or by running them off." They argued that the farmers could not fence their fields at present "because it is almost time for the harvest and there is so much fencing to be done." They insisted that the soldiers should hire a herder because they were each paid 380 pesos a year by the king, whereas the farmers had "no source of livelihood other than the small fields." The farmers also complained that they could not drive the cattle because the cattle had gored and killed their horses when the farmers tried to drive them out. The farmers claimed that they had never killed "a single head of their cattle" and that the cattle owners owed them for more than four hundred *fanegas* (one fanega equals 2.58 bushels) of corn that the livestock had eaten. However, the governor rejected the cabildo's pleas, ordering the farmers to build the fence and suggesting that future petitions "be drafted with the wisdom and prudence conducive to justice."[17]

Disputes among the isleño farmers, the soldiers, and the missionaries continued from the 1740s through the 1770s with charges, countercharges, and decrees that were regularly ignored. The stock raisers were ordered to use caretakers to keep their animals away from crops, and farmers were admonished to fence their fields. When caretakers could not be provided, stock raisers were to keep cattle a reasonable distance from cultivated fields. Considering that the Apaches and Comanches were often threats during this period, it is not surprising that the herds were never brought under satisfactory control.[18]

In 1762 several residents of the villa complained that because they owned no land, they were "totally helpless, deprived of the benefit of being able to advance and cultivate the lands." They petitioned the governor to examine "the possible cost of the withdrawal . . . of water" from the river and distribution to landless citizens. As a result, the governor appointed Geronimo Flores, a citizen he described as "skillful in withdrawing water," to "measure the lands which said withdrawal might irrigate, . . . estimate the difficulties which present themselves concerning the dam and canal and estimate what everything would cost." Within a short time Flores accomplished his task and reported that if water were taken from the river about six thousand varas north of the villa, an acequia could irrigate land north of the villa between San Pedro Creek and the river. But the cost of the dam and acequia would be at least three thousand pesos, "even with the assistance of the personal labor of the community." The governor forwarded Flores's report to the viceroy; however, the Spanish bureaucracy moved slowly, and work on the dam and acequia did not begin until about 1776, after the governor had determined that there was no legal reason why a new irrigation system could not take water from the river above the town as long as sufficient water remained to irrigate the mission fields downstream. A number of families contributed labor or other resources to the effort and received irrigated land and water rights in return. The land, called the *labor de arriba,* and associated water rights were distributed in two drawings, one in 1777 and a second in 1778. When completed, the system watered about six hundred acres north of the villa between the river and San Pedro Creek (see Map 2).[19]

FARMING PRACTICES

During the 1740s progress was rapid at the San Antonio missions. By 1745 a total of 885 Indians lived at Valero, Concepción, San Juan, and Espada. All the San Antonio missions had irrigation systems and were raising good crops. That year the corn crop was eight thousand bushels, about 1.4 pounds per day for each Indian resident. If we assume that corn yields averaged only twenty bushels per acre, about four hundred acres would have been needed to produce the crop. This was less than one-fourth of the land that could be irrigated, suggesting that the missions left fields fallow for several years between crops so they could regain their fertility. In addition, surpluses of some vegetables were produced. Cotton was cultivated at Valero and sometimes at Espada. Together, Valero, Concepción, San Juan, and Espada owned over 5,100 head of cattle, 2,600 sheep, 660 goats, and 250 horses.[20]

The cropping cycle in the San Antonio valley was similar for both mission and civilian farmers. Work normally began in January with repair and maintenance of the irrigation ditches and fences. If the weather was dry, irrigation could be used to soften the soil for tillage and leave a moist seedbed for planting. The soil was prepared for planting with teams of

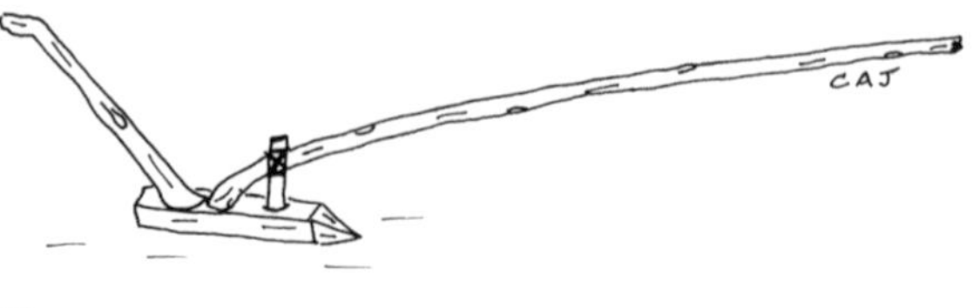

Plow

oxen pulling simple wooden plows. The main body of the plow usually consisted of a limb or the trunk of a small tree with one end sharpened and covered with a piece of iron to minimize wear. A branch growing from or attached to the rear of the body served as a handle to guide the plow. A pair of oxen was lashed to a long pole or beam attached to the body near the base of the handle. The beam was stabilized by a stout *tranca,* or vertical support, firmly attached to the body of the plow in front of the handle. Two farmers tended each plow, one to guide it and another to goad the oxen. In order to plow and plant the mission fields, several teams worked together, and several passes with these primitive plows must have been required to prepare the soil for planting.[21]

In March, sugarcane, beans, peppers, and the first crop of corn were planted. Vegetable gardens were planted in March and April. The late crop of corn and more beans were sown in May and early June. The first harvests of beans, peppers, and early corn came at the end of June and in early July. Other harvests followed throughout the summer and into the autumn. The last beans were typically harvested in November, and mature corn could be left in the field until December or January.[22]

After harvest, the grain and vegetables were stored in the missions' granaries. Throughout the year, food was distributed weekly or in some cases, daily, to the Indians. If food was in short supply, the missionaries used a part of their budget to purchase it and other supplies for the Indians. When a surplus was available, the missionaries, as representatives of the Indians, sold food to the presidios, nearby settlers, or sometimes to other missions.[23]

Because the settlements along the San Antonio had to be almost self-sufficient in food production, failure of the corn crop could have serious effects. In June 1738 an "uncommon plague of locusts [grasshoppers]" caused twelve of the villa's most prominent citizens to appeal to the governor for assistance. The grasshoppers had ruined their summer crops, and there were no seeds for planting cool-season crops that would escape the plague. They requested the governor to send them at least nine fanegas of wheat; one fanega each of chickpeas, beans, and peas; and half a fanega each of lentils and green peas. They promised to pay the cost of the seeds, "but the manner and form of payment must be contingent on the yield of the land, for it should be understood that we cannot promise cash because we do not have any." The governor, recognizing the seriousness of the situation, ordered the seeds from Saltillo.[24]

SUCCESS OF THE MISSIONS

Clashes with the Apaches continued throughout the 1730s, but they began to decrease in the 1740s. Despite attacks by the Spaniards on rancherias northwest of San Antonio in the spring of 1745, a retaliatory attack on the presidio in June, and additional hostilities in 1748 and 1749, the Apaches expressed a sincere desire for peace. In August 1749, pressured by the Comanches, who had begun to encroach on their hunting grounds from the north, the Apaches became allies of the Spaniards. As hostilities declined, the missions began to expand their herds and utilize the abundant prairies stretching for miles on both sides of the river. With peace, it became easier for soldiers and vecinos to hunt mesteño cattle for their own consumption. But the major livestock owners had always been the missions, and the padres objected. In response, in 1751 the alcalde prohibited anyone but resident stock owners, effectively the missions, from killing cattle within thirty leagues of San Antonio, punishable by a fine of twenty-five gold pesos for a Spaniard or two hundred lashes and banishment for a mestizo or mulatto.[25]

During the 1750s and 1760s the mission herds of cattle and horses remained relatively stable. In 1762 the five San Antonio missions claimed almost 5,500 cattle, 560 saddle horses, 15,200 sheep and goats, and about 1,200 brood mares. A few years later, in 1768, livestock at the missions totaled 5,487 cattle, over 600 saddle horses, almost 1,000 mares, more than 1,000 donkeys and almost as many mules, and 17,000 sheep and goats. It is noteworthy that in 1745 and 1768 the number of cattle claimed by the San Antonio missions changed little; however, the number of sheep and goats increased from a little over half to about three times the number of cattle. Perhaps peace with the Apaches permitted mission sheepherders to provide better care and protection for their flocks.[26]

San José was the most beautiful and best organized of the San Antonio missions. In 1762 a total of 350 Indians from eight tribes were in residence. With the exception of new arrivals, most spoke Spanish, played a musical instrument, and performed Spanish dances. Those who failed to perform their assigned tasks or attend religious services were tried and punished by Indian officials. No Spanish overseers were needed to

Mission San José

manage the work in the fields and workshops. Able-bodied Indian men planted and harvested the crops, cared for the livestock, managed the carpentry shop and forge, quarried stone, burned lime for mortar, and made adobe and roof tiles. Girls and women spun wool and cotton, wove cloth, and sewed—enough to provide each person with two sets of clothing, one for work and the other for feast days. Old men made arrows, old women caught fish, and young boys and girls went to school and were taught to recite prayers. The missionaries of San José lived and worked in a stone friary containing their cells, offices, a kitchen, and a refectory. The mission's stone granary could hold twenty-five hundred bushels of corn and beans, and thirty yokes of oxen were available to plow the fields and transport materials. Cultivated fields were all fenced and were irrigated by the San José acequia, in which the Indians fished. The canal also passed through two swimming pools, one for the Indians and the other for the soldiers. Crops included corn, beans, lentils, vegetables, chilies, melons, potatoes, and sugarcane. A small sugar mill, the first in Texas, made *piloncillo* (small cones of raw sugar), and orchards produced peaches weighing a pound apiece. The agricultural production supplied all the needs of San José, and the excess provided food for the presidio and the settlements at La Bahía, Orcoquisac, and Los Adaes. Ten or twelve leagues away, the mission had a ranch called Atascosito where in 1767 it pastured ten *manadas* of brood mares (each manada contained about twenty mares), almost 100 saddle horses, four manadas of burros, and about 5,000 sheep and goats; however, the mission had only about 1,500 branded cattle, having lost 2,000 in recent years to Indian raids.[27]

Valero was also prosperous. Though its bell tower and sacristy had collapsed, a new church was planned. The mission's stone wall was strong, and the missionaries lived in a two-story *convento* containing their cells, offices, a kitchen, workshops, and a granary. In 1762 it sheltered 275 Indians from seven tribes. They now lived in stone houses with doors, windows, and arched porticoes, all arranged around a square. They were furnished with beds, chests with drawers, *metates* (grindstones), *comales* (flat irons to cook tortillas), pots, pans, and other utensils. Willows and fruit trees grew along the irrigation ditch that ran through the compound, and a well had been dug to supply the inhabitants in case of siege. Crops grown at Valero included corn, beans, chilies, cotton, and vegetables, which were worked with forty yokes of oxen, thirty plows, a number of harrows, forty hoes, and twenty-five scythes. Twelve carts were used to transport crops, tools, supplies, stone, and timber. The Indians also manufactured a variety of cotton and woolen fabrics in workshops containing four looms and two storerooms. The stone granary could store eighteen hundred bushels of corn, about one pound for each inhabitant for each day of the year, and several hundred bushels of beans. There was also a well-equipped blacksmith shop. Valero's ranch had 115 saddle horses, 1,115 head of cattle, 2,300 sheep and goats, 200 brood mares, 15 donkeys, and 18 mules. A three-room rock house twenty-five varas long was provided for ranch workers.[28]

By the 1760s the other three San Antonio missions also had substantial stone buildings, though at San Juan and Concepción, Indians still lived, at least in part, in thatched houses. Agriculture at the other San Antonio missions had also improved since 1745. Their fenced and irrigated fields allowed corn, beans, cotton, chilies, and vegetables to be grown in abundance, and each mission had at least sixteen hundred bushels of corn still on hand in the spring of 1762, before the summer harvest.[29]

Throughout the 1760s abundant food and the relative safety of mission life helped attract neophytes from the small Coahuiltecan bands. Caught between the marauding Plains tribes and the Spanish missionaries seeking converts, many Coahuiltecans chose the well-fed, though regimented, life of the missions. However, they sometimes became dissatisfied, and it was not unusual for two or three families to steal away at night and travel continuously well into the next day. When they were missed at the mandatory morning service, a missionary would set off in pursuit, often with one or two soldiers. When finally encountered, the

Indians sometimes returned willingly to the mission; otherwise, they were recaptured. Pagés reported that the soldiers "use the thong [rope], and lace [lasso] them like wild horses." Thus caught, an Indian captive was "bound hand and foot, and carried to the residency of the missionary, who makes it his business, by threats, persuasion, severe fasting, gentleness, and last of all by marriage, to tame and civilize the manners of his prisoner."[30]

Just as the missions in the San Antonio area were becoming prosperous, new settlements were growing thirty-two leagues down the Río San Antonio near present-day Goliad. In 1726 Mission Espíritu Santo and Presidio Nuestra Señora de Loreto (also known as Presidio La Bahía) were moved to the Guadalupe River near present-day Victoria, and in 1749 they moved again to their final location on the Río San Antonio. Along with Mission Rosario, founded in 1754, these institutions were established to Christianize the Karankawa Indians, a daunting task. Presidio La Bahía was built on a hill, and its wall, constructed of wooden beams, river rock, and lime mortar, enclosed an area 350 by 370 feet. Mission Espíritu Santo housed more than two hundred Indians in 1762. It had a fine stone church, a monastery with closed cloisters, Indian houses made of stone, and a cemetery with a chapel. In 1758 its neophytes claimed 3,200 cattle, 1,600 sheep, and 120 horses. They harvested more than one thousand bushels of corn a year from its unirrigated fields and produced cotton and wool that were woven into fabric at the mission. Espíritu Santo's enormous ranch, like the mission, was on the east side of the Río San Antonio and extended to the Guadalupe River.

Located about four miles west of the presidio, Rosario had an enclosed compound about 300 feet on each side. The initial buildings were of wood covered with mud and plastered with lime. Later, a new church, a granary, and housing for the missionaries were built of a mixture of rock, gravel, and clay. Indian huts were of adobe and rock. Rosario's large ranch, like the mission, was on the river's west bank, and it extended far to the northwest. By 1758 the four hundred neophytes claimed 700 cattle, 150 sheep, and 50 horses. Despite a lack of irrigation, Rosario's fields grew corn, beans, potatoes, and sugarcane; and it produced peaches, apples, and figs in its orchards. But crop production at the La Bahía missions was unreliable. As a result, regular supply trains, each escorted by a squad of seven soldiers, brought the necessary corn, beans, and vegetables from the San Antonio missions. Nevertheless, the herds at La Bahía increased rapidly in the 1760s and 1770s, when over 40,000 cattle, mostly unbranded, roamed the missions' extensive pastures.[31]

TRAVEL AND TRADE

The community of San Antonio was blessed with abundant irrigation water, fertile soils, and almost limitless pastures. Though the mission Indians and vecinos had the means to produce an abundance of agricultural products, they had limited capacity to consume or trade them. Long distances, marauding Indians, extreme weather, poor roads, dangerous rivers, and legal obstacles inhibited trade with both Louisiana and the interior of Mexico. The difficulty and cost of reaching export markets limited the amount of money in the community, depressed economic activity, limited imports, and discouraged agricultural production.

Travel and transportation of goods were fraught with difficulties and dangers. In 1767 Pierre Pagés described the road between eastern Texas and San Antonio as "difficult to be found, and across rivers, many of which are extremely dangerous in their passage." He noted that "it would be deemed highly imprudent to attempt it with fewer than ten or twelve persons in company." The party he accompanied from Nacogdoches included fifteen "soldiers and half-savages," twenty loaded mules, and "a reserve of no less than two hundred horses." At the Trinity River, which was running "with a strong current at least two gunshots in breadth," they crossed in columns with their packhorses in the middle and riders on horseback on each side to break the force of the current. The Brazos was judged to be even more treacherous; therefore, large trees were cut, rolled to the river, and tied together with the halters of the horses to make a raft. The party's baggage was then placed on the raft, and guided by two experienced swimmers in front and one on each side, the raft was "transported, or rather shot across the stream." The horses and mules were then driven to the edge of the river and led across "by a savage on horseback" while others "who stood on the opposite bank, were at much pains, by shouting and clamour, to attract their notice, and to entice them to that particular spot."[32]

The roads leading southwest were equally dangerous. In February 1768, Fray José de Solís traveled from Zacatecas to Texas. After crossing the Río Grande near present-day Laredo, he was met by an escort consisting of a missionary, eight soldiers, and four Indians armed with guns "because the road which I was now to travel was infested with unfriendly Indians." Water was almost always scarce, and travelers were careful to stop at known campsites where good water could usually be found.[33]

Most of the trade goods that moved between San Antonio and towns to the east and south were transported on pack mules driven by *arrieros*, or mule drivers. A day's march for a mule train was ordinarily fifteen to twenty miles, and the trip from Laredo to San Antonio took ten to fifteen days, though "military couriers, furnished with remounts and traveling part of the night" could complete the journey in only two or three days. In September 1777 a mule train traveling from Presidio del Río Grande to San Antonio was attacked by Apaches at night near present-day Cotulla. Two of the nine travelers tried to resist and were killed. The others saved themselves by fleeing into the brush. Over fifty years later, in 1828, Jean Louis Berlandier wrote that the road from Laredo to San Antonio still was "not very safe; the Apaches and Comanches infest it at every step."[34]

The men who entered this dangerous and risky business, living for weeks at a time on the road exposed to the elements and hostile Indians, often had little regard for law and order. The arrieros, usually illiterate and often solitary men, left little direct record of their business. However, they were not infrequently accused of fraud or smuggling, resulting in legal proceedings that included inventories of their possessions. For example, a suit in 1733 revealed that arriero Josepho Antoni Rodrigues owned fifteen gentle pack mules, nine packsaddles, six gentle horses, a saddle, several saddle pads, a new bridle, and associated equipment. Trade goods in his possession included eight boxes of piloncillo from Huasteca (each containing 160 of the sugar candies), six *arrobas* (one arroba equals twenty-five libras or 25.4 pounds) of salt from Colima, and twenty strings of *chile ancho* from Colima (each string twelve and one-fourth varas long). For his next trip south, he had accumulated ten bison hides and thirty deer hides, all tanned with the hair left on. In 1750 arriero Juan Joseph Sevallos was sued for abducting a slave woman. The inventory of his property revealed that he owned twenty-three mules (twenty with packsaddles and other trappings), three horses, three burlap sacks of meal, one box of piloncillo candy, some chilies packed in burlap, thirteen loads of burlap sacks, one barrel of quasi-commercial wine mixed with must, and one barrel of good commercial *aguardiente,* a distilled liquor made from grapes or sugar cane. It is interesting that Antoni and Sevallos brought piloncillo and chilies to San Antonio, when both were produced at the missions.[35]

As the community on the Río San Antonio grew and matured, the missions and civilian populace developed an efficient system of irrigated agriculture. However, as the population of mission Indians declined after the 1760s, the community found itself with more irrigated land than it needed to produce the crops for local consumption. With only limited markets for its crops, it became clear that cattle were a potential source of wealth that could be enjoyed by the civilian as well as the mission population. Cattle could be slaughtered for meat, tallow, and hides, all of which could be sold in San Antonio, Saltillo, or other towns on the northern frontier. The ranching industry was also a potential source of tax revenue for the government. But development of the industry required coordination among the mission ranches, civilian ranchers, and government officials, as well as relative peace with the Apaches and Comanches—a combination difficult to achieve.

Chapter 3

RANCHING ALONG THE RÍO SAN ANTONIO

THE EARLY PRIVATE RANCHING FAMILIES along the Río San Antonio were, like their counterparts all across the frontier of New Spain, descendants of soldiers. Rewarded for their service with grants of land, officers often became part-time stock raisers, taking up ranching full-time when they retired. The first privately owned ranches in the San Antonio area were Andrés Hernández's San Bartolomé and Luis Antonio Menchaca's San Francisco. The two ranches were legalized in 1758 in the settlement of a dispute over ownership. According to court documents, Hernandez's family had been in possession of San Bartolomé since 1718. In 1737 the family's claim had been ratified by the governor, and Hernández's father had occupied the ranch in the early 1750s. But Menchaca, who was then captain of the presidio, challenged Hernandez's claim and forced a compromise in which Hernández lost more than half of his land. Located in El Rincón, the corner or fork formed by the confluence of the Río San Antonio and Cíbolo Creek, San Bartolomé consisted of four *leguas* and eight *caballerías*. San Francisco contained eleven leguas and four caballerías.[1] By modern standards these were large ranches. One legua, called a league in English, is equivalent to five thousand varas, about 2.6 miles. When used as a unit of land, it is an area one legua square, or about 4,428 acres. This area is also known as a *sitio de ganado mayor.* One caballería was the amount of land needed to pasture one horse for service to the Crown, about 105.8 acres.

Throughout New Spain, including Texas, the government allowed any citizen to request a land grant. The petitioner usually emphasized past service to the Crown. If successful, the grantee was required to pay survey and title fees, occupy and develop the land, and pledge to defend both the land and the province in times of danger. Of course, the missionaries, who claimed that all the cattle in the region were descended from their original herds, opposed establishment of private ranches. In 1756 seven respected members of the cabildo, including Menchaca, complained to the captain of the presidio and his superiors that as soon as the missionaries learned that vecinos were seeking a grant on the Río Guadalupe north of San Antonio, they constructed *jacales,* simple wood and thatch huts, and moved cattle and sheep onto the land. Menchaca and his colleagues complained that it was "not necessary for the missions to monopolize the land because they have so few Indians." In fact, if the missionaries did not bring Indians "by force from the coast, there would not be any Indians in the missions." The members of the cabildo feared that "these lands will never be turned over to the common people."[2]

Little by little the government responded to ranchers' requests for land. For example, in 1766 Domingo Castelo convinced a magistrate to order Mission San José to sell him eleven leagues of land near present-day Castroville for 130 pesos. The number of private ranches continued to increase during the 1760s and 1770s, and historians have been able to identify and locate twenty-four specific ranches belonging either to the missions or to private individuals in about 1776. By 1780 about forty ranches can be identified. In addition, the names of at least sixty-four cattle ranchers can be found in historical records from 1778 to 1782.[3]

In 1767 Pierre Pagés reported that the ranchers' "principal employment is to rear horses, mules, cows, and sheep." However, because of limited legal markets and continued harassment by the Comanches and their allies, ranchers along the San Antonio neglected their ever-increasing herds. Pagés reported that as a result, "their horses and mules are no sooner a little broken in, than they are offered to sale; but here the market price is so extremely low . . . that I have seen a good horse sold for a pair of shoes." In addition, "horned cattle, which were originally tame, but have long since become wild, . . . now roam in large herds all over the plains" east of San Antonio. About every two months some ranchers rounded up their cattle. Described by Pagés as "hunting the wild bull," the roundup began "with a sort of festivity." When the vaqueros

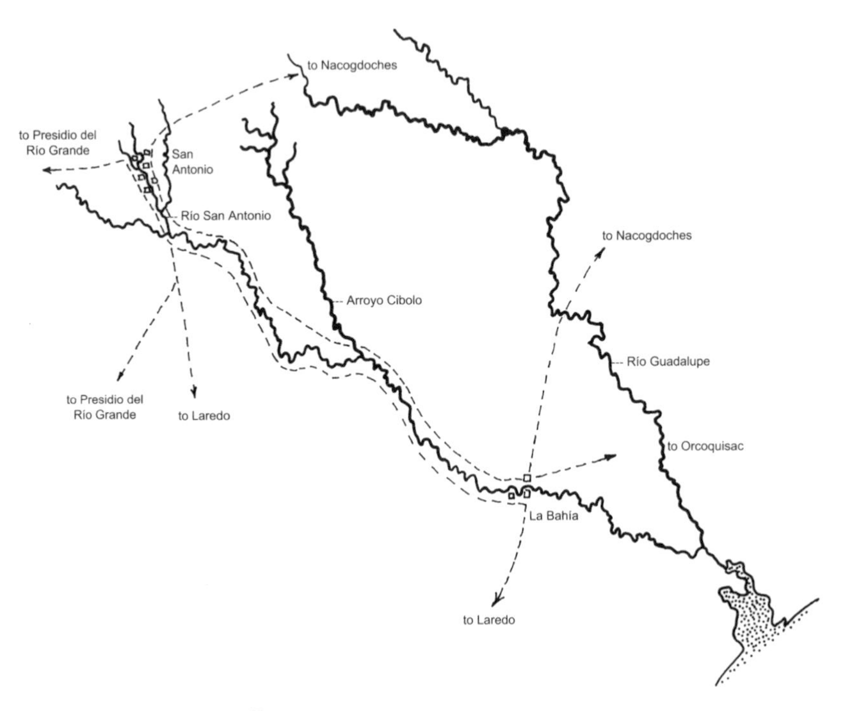

MAP 3. *San Antonio Valley.*

discovered a herd of cattle, they formed themselves into a crescent and drove the animals until "becoming jaded with fatigue, they were ready to sink under the whip." Animals that were unable to continue "were slaughtered on the spot," and when the vaqueros had "supplied themselves with what quantity of their flesh was deemed sufficient, [they] abandoned the rest to the wolves." The tired cattle that remained were driven "into fields adjoining to the houses of their owners, where every means is used to render them tame and tractable." After being "subjected to hunger and confinement, they receive their liberty, and are succeeded by others, which experience in their turn a similar course of discipline."[4]

San Antonio's "excellent horsemen," were "particularly fond of hunting or *lacing* [lassoing] their wild animals." As soon as his horse was within "a certain distance" of the cow or horse being pursued, the vaquero would throw "the running noose . . . with such dexterity, that he seldom fails to catch him round the neck or horns; and in the same instant, by pulling up his horse, or turning him abruptly from the line of his career, he checks his prey, and obliges him to stand still."[5]

Spanish stock raisers used several terms to denote different states of wildness and ownership. *Ganado vacuno* and *manso* referred to gentle animals that could be easily managed. *Orejano* meant an animal that had no mark cut into its ear (*oreja*); thus, it was unbranded and unmarked. *Alzado* referred to a "stray" that had wandered or run wild, though the animal might have a recognizable brand. A *mostrenco* was an animal with no known owner, and when it was found, it had to be exhibited to allow the rightful owner to claim it. A *cimarron* was an animal that was too wild to be tamed or handled effectively. The term *mesteño* meant an alzado, orejano, or mostrenco animal and could refer to either cattle or horses.[6]

TROUBLES ON THE RANGE

In 1770 a new governor, Juan María, Barón de Ripperdá, established his headquarters in San Antonio. A loyal, energetic, and morally rigid Spaniard, Ripperdá was a strong advocate for the Crown and the Church; however, his energy was often directed toward prosecuting the vices common in the villa. This soon brought him into conflict with some of the leading citizens of San Antonio, many of whom were ranchers with flexible notions of cattle ownership. At the core of Governor Ripperdá's disputes with the ranchers was his desire to protect the mission herds from capture and sale by private ranchers. Accepting the missionaries' claims that they were the first ranchers and all unbranded cattle could be assumed to be the offspring of their herds, Ripperdá did not hesitate to prosecute private ranchers whom he suspected of capturing or killing mission cattle. In 1774 he seized some cattle that rancher Carlos Martinez attempted to sell in San Antonio, alleging that they were from the missions' herds. In 1774 and 1775 he prosecuted meat hunters for killing cattle on mission ranches, even though they had been granted permission to kill cattle for meat on lands beyond the mission ranches.[7]

In March 1777 Governor Ripperdá fined and jailed Marcos de Castro, Miguel Guerra, and Guerra's son Francisco for "seizing herds of unbranded cattle" from pastures belonging to the missions and placing their own brands on them. These proceedings unleashed a long series of depositions, accusations, petitions, and confessions by private ranchers throughout 1777. The ranchers claimed that they should be allowed to capture, brand, and export unmarked cattle roaming the vast pastures that had long been considered property of the missions. In December 1777 Ripperdá found that several private ranchers "customarily ran and took unbranded cattle on others' [the missions'] summer pastures." Ripperdá also established that in 1776 ranchers had gathered a large herd of unbranded cattle and had driven them south and sold them "at the ranches of Camargo and Laredo, Vallesillo, Boca de Leones, and other towns." These proceedings undoubtedly increased animosity toward the governor, who throughout his term favored the missions and repeatedly attempted to correct moral weaknesses of San Antonio's citizens.[8]

Regardless of what the ranchers thought of Ripperdá at the end of 1777, they could hardly have expected the changes that would occur in January 1778, when Teodoro de Croix, governor and commandant general of the interior provinces of New Spain, arrived in San Antonio. Vested by the king with "the same faculties and duties as those formerly held by the Viceroys" in Mexico City, Croix was visiting Texas to oversee "the upright administration of justice in affairs both civil and criminal" in order that "scandalous public iniquities shall cease, as well as enmities, factions, and disputes among citizens, or among citizens and their judges." With this laudatory goal in mind, Croix pronounced that Texas' "backwardness" and lack of progress were due primarily to "laxity of conduct," including "theft, incontinence, scandalous concubinage, prohibited games and drinking, and the lenience with which many vicious, slothful, idle, and vagabond persons have been allowed to establish themselves."[9]

Father Agustín Morfí, who accompanied Croix to Texas, noted that cattle were the people's "greatest refuge," but the ranchers were "clandestinely killing" them "to sell [tallow and dried meat] to the presidios and settlements of Coahuila and Nuevo León." He also reported that the private ranchers were "in perpetual controversy with the missionaries and governors." Croix declared that as a result of these abuses and conflicts the herds of horses and cattle around San Antonio had "diminished considerably." In addition, even though Governor Ripperdá had attempted to stop them, "everyone, up to now, has considered himself authorized to go out and build corrals or stockades, then round up, enclose and take possession of wild and unbranded cattle and horses."[10]

Croix therefore declared that all unbranded cattle and horses now belonged to the king, "first because they are strays and have no known owner, and also because they are born and raised in his unappropriated lands." To govern the ownership and export of livestock, Croix established the following rules:

- No person would be allowed to "go out and round up, kill, or take wild or unbranded cattle or horses."
- No one would be permitted to "take from the province any herd of cattle or horses, even if

they are domestic, branded, and of his own breeding stock, without first obtaining for that purpose a [free] license from the magistrate."

- To prevent unauthorized branding of wild cattle and horses or unbranded offspring of the missions' herds, all livestock owners, including the missions would be allowed "to brand only those times designated as suitable by the magistrates, and with their attendance or the attendance of some person commissioned for that purpose."
- The governor would grant licenses and collect fees (four *reales* for cattle, six for horses; eight reales equal one peso) for capturing "wild and unbranded cattle or horses in the times, seasons, and sites to be designated."

Croix's proclamation spelled out in detail the fines associated with breaking his rules. In addition, he ordered that records of export licenses, livestock branded, and fees and fines collected should be kept in the "mesteño fund" under "triple lock and key" with one key entrusted to the governor, another to the highest-ranking alcalde, and the third to the oldest regidor.[11]

Both the missionaries and the ranchers protested Croix's decree, each asserting that the unbranded cattle roaming the prairies were the offspring of its herds. The missionaries complained that because the government could not protect them from hostile Indians, the neophytes had been unable to brand their cattle. In response, Croix granted the missions a grace period of six months; however, the Comanches had stolen all but a few of the mission horses, making it impossible to gather and brand the cattle. Despite their protests, some private ranchers began to comply, and by November 1778 San Antonio cattle ranchers had legally exported a herd of 100 cows and bulls to Louisiana, a herd of 779 "to market in the frontier presidios," and herds of 311 and 100 to unspecified destinations. In addition, records indicate that 353 calves and older cattle were branded for sixty- two owners.[12]

Athanase de Mézièrs, a Frenchman from Natchitoches chosen to help Croix put his decree into effect, reported large numbers of wild cattle and horses between eastern Texas and San Antonio and reasonable prices for all classes of livestock. Fat cows sold for four pesos, sheep three and a half years old for six pesos, breeding ewes and goats for three pesos, half-broken horses for six pesos, manadas of mares for one peso per animal, and wild mules for eight pesos. The cheapest cattle could be obtained at Mission Espíritu Santo. The best season to buy and drive herds to Louisiana was the fall, when the rivers were low, the pastures were good, and the trip from the valley of the San Antonio to Opelousas took only about a month.[13]

For many years Mission Espíritu Santo had been unable to brand the cattle on its enormous ranches east of the Río San Antonio. These wild, unmarked herds were now property of the king and could be caught and branded for a small fee. Nevertheless, a group of ranchers, probably hoping to take advantage of the confusion caused by Croix's decree, began secretly to gather and brand cattle on Espíritu Santo's ranch. But Ripperdá soon learned of their activities. After taking dozens of depositions from the ranchers and the vaqueros they employed, he concluded that during the four-month period after Croix's decree eight prominent ranchers conducted eight roundups in which they "ran, stole, killed, marked, and branded more than four hundred head of unbranded cattle" from Espíritu Santo's pastures. Unable to complete the case before he left San Antonio, Ripperdá turned the proceedings over to his replacement, Domingo Cabello, on November 1, 1778. Cabello was immediately embroiled in appeals of Ripperdá's rulings, as well as other cases involving disputed ownership of cattle.[14]

By 1778 several ranchers had begun to accumulate substantial herds. Luis Antonio Menchaca, the province's wealthiest citizen, owned 2,444 head of livestock (including cattle, horses, mules, sheep, and goats), almost one-third the total 7,434 head owned by 194 vecinos, and more than twice the 1,067 (probably mostly horses) owned by the 79 soldiers at the presidio in San Antonio. In 1779 in Béxar, 140 civilian households owned 3,035 cattle; however, over 76 percent of those belonged to only 10 owners. Menchaca owned 850 head, followed by María Ana Curbelo with 300, Simón de Arocha with 250, and Macario Zambrano with 230. In addition to cattle, the civilian community also owned 207 yokes of oxen, 520 horses, 499 mares, 241 mules, 36 donkeys, and 1,665 sheep and goats. Whereas more than 153 households owned horses and 140 owned cattle, only 6 owned sheep and goats, and 3 owned donkeys.[15]

RODEOS AND EXPORTS

Cattle had little respect for ranch boundaries, and livestock from several ranches often mingled on the prairies around San Antonio and La Bahía.

To prevent the herds from wandering too far from home, ranchers routinely gathered their cattle and kept them for a few days in corrals near the ranch headquarters. But once or twice a year, usually from October to December, several ranches might join together to conduct large organized roundups, or rodeos. A *jues de campo,* or field judge, was appointed to oversee and manage the roundup and branding. On the appointed day, vaqueros from all the participating ranches and missions rode out, formed a large circle, and drove the cattle toward a central point. The first major roundup after Croix's decree occurred in 1779, and the difficulties encountered were indicative of the conflicts among the missionaries, independent ranchers, and government officials. Most of the cattle in the province were found around La Bahía, but Indian raids and a shortage of herders had prevented both the missions and independent ranches from regularly rounding up and branding their stock. As a result, the herds between San Antonio and La Bahía were mixed and contained many adult unmarked (orejano) cattle. Because Croix had not decided the ownership of the orejanos, the governor would not permit them to be branded. The ranchers and missionaries of Espíritu Santo were charged with rounding up all the cattle carrying their brands, branding the calves following their cows, separating the herds of different owners, and returning them to their rightful pastures. But almost all of the mission's Indian neophytes had fled, and it lacked the vaqueros to participate. As a result, the governor permitted the mission to appoint an agent to assure that its orejanos and calves were cut out before branding began.[16]

In early November 1780 Governor Cabello issued orders for the annual rodeo and branding. In order to make the roundup as orderly and efficient as possible, he divided the participating ranches into three groups, each of which would gather and brand cattle in a specific area. The three groups consisted of the ranches along the San Antonio, those along Cíbolo Creek, and Rancho Pataguilla, which would conduct its own rodeo. Five ranches did not participate because of poor weather; however, seven ranches rounded up a total of 3,454 branded cattle and branded 921 calves. In the 1781 roundup, which was actually conducted in January 1782, ten ranches gathered a total of 3,613 head and branded 1,964. The large number branded reflects Cabello's order that unmarked animals less than three years old could be branded, presumably because it had not been possible to brand all the calves the previous two years. Valero was the only mission that took part in the roundup.[17]

In response to continued claims by private ranchers that they owned the unbranded livestock that roamed their properties, in 1783 Governor

Cabello issued a new edict reaffirming the Croix proclamation of 1778. However, in recognition of the ranchers' complaints, he gave them four months, and after appeal by the missions, an additional two months "to round up all the livestock they own and to place thereon their known brands and marks." Thereafter, any unbranded livestock (with the exception of young calves) on any mission or private ranch would be considered government property, subject to the rules originally stated by Croix in 1778. The "orderly and methodical catching of orejano cattle for the purpose of increasing haciendas, taking them out [of the province], or slaughtering them" was permitted, but only if the rancher obtained permission from the governor and paid the required fee: twenty reales for cattle at least three years old, sixteen reales for two-year-olds, and eight reales for yearlings (one peso equals eight reales). Sales of this kind required the herd to be inspected by a representative of the governor, and local ranchers were given preference over outsiders.[18]

During Governor Cabello's term, from 1779 to 1786, ranchers around San Antonio legally exported more than 18,000 cattle to Louisiana, Coahuila, and Nuevo León. Of this total, only 2,434 were exported to Louisiana, in 1780 and 1782. The totals legally exported ranged from a low of 1,079 in 1779 to 3,374 in 1786. In a given year the number of participating ranchers varied from three in 1780 to eleven in 1786.[19]

Records indicate that during 1784 the mesteño fund received the following deposits:

- 140 pesos 4 reales from Manuel de Arocha for 75 orejano cattle exported to Coahuila
- 852 pesos from Luís Mariano Menchaca for 410 orejano cattle exported to Coahuila
- 333 pesos from Santiago Seguín for 177 orejano cattle exported to Saltillo
- 375 pesos from Juan José Pacheco for 250 orejano cattle exported to Saltillo
- 67 pesos from Luís Mariano Menchaca for 144 cattle purchased from Nicolás de la Mathé
- 265 pesos 4 reales from Francisco Xavier Rodriguez for 139 orejano cattle exported to Saltillo
- 501 pesos from other soldiers and residents of San Antonio, probably for capture of an unspecified number of orejano cattle and mesteño horses

- 51 pesos from soldiers and residents of La Bahía, probably for capture of an unspecified number of orejano cattle and mesteño horses
- 70 pesos 2 reales from Antonio Gil Ybarbo for an unspecified number of orejano cattle and mesteño horses captured near Nacogdoches

In addition, Menchaca had apparently captured and exported 1,000 orejano cattle for which he had not paid, due to ongoing legal proceedings. Mission Espíritu Santo owed the fund 957 pesos that Cabello had been unable to collect because the amount was "still being discussed."[20]

Ranchers who exported cattle often drove herds of both orejanos and branded cattle, some purchased from or driven for their owners. For example, in July 1784 Santiago Seguín applied for and received permission from Governor Cabello to export 177 orejano cattle that he had gathered on Mission Refugio's ranch. He was charged an export fee of 333 pesos for the cattle: 81 cows three years or older (20 reales each), 30 two-year-old cows (16 reales each), 25 two-year-old bulls (8 reales each), 16 yearling heifers (8 reales each), and 8 yearling bulls (4 reales each). In addition, Seguín was authorized to export 93 cattle "branded with the brands of various residents and these settlements so that he may drive them to wherever he deems most suitable."[21]

Despite the apparent success of his policies, in 1785 Governor Cabello called attention to the difficulty of preventing fraud and collecting the prescribed taxes. He complained that livestock were "maintained in fields so exceedingly sparse and open that anyone who wishes may commit frauds, which are restrained only by my zeal, vigilance, and efficiency." An example of fraud occurred in May 1784 when Luis Mariano Menchaca, a member of the villa's cabildo, requested Cabello's permission to export to Coahuila "a herd of orejano cattle, both cows and bulls." The license was granted, and Menchaca rounded up 126 bulls, 138 cows aged three years or older, 146 two-year-old cows, and 52 calves, paying a total of 852 pesos. However, Cabello soon received "a secret denunciation" stating that in addition to the cattle Menchaca had exported, he had captured others, "placed his mark on them," and kept them on his ranch without paying the required fees. Witnesses, including vaqueros who took part in gathering and branding the cattle, testified that in addition to the 462 head that were legally exported to Coahuila, about 200 head were branded and left on Menchaca's ranch. Cabello declared that this was another instance of "the inveterate custom of the residents of these settlements" to appropriate "as many orejano cattle as they can take without paying the due royal fees." Menchaca appealed to the commandant general that Cabello's proceedings were malicious and based on false accusations, and the unresolved suit was turned over to the new governor, Rafael Martínez Pacheco, in November 1786.[22]

Another example of corruption occurred in July 1784. Fray José Francisco Mariano de Cárdenas of Mission Espíritu Santo requested permission from Cabello for Blas María de Ecay y Músquiz to export cattle he had purchased from the mission. Cabello replied that he would grant permission to export only branded cattle, and he wrote a letter to Captain Luis Cazorla at La Bahía asking him to grant an export license to Ecay for 1,000 branded cattle belonging to the mission. Cazorla replied, requesting permission on behalf of Fray Mariano to export some orejanos if 1,000 branded cattle could not be found. Cabello gave his consent on condition that full export fees be paid to the mesteño fund, even though he felt sure that at least 2,000 branded cattle roamed the mission's ranchlands. By early September the cattle had been gathered, and Cazorla reported that 530 cows and 47 bulls and bullocks, all branded, had been gathered for export. A few days later, Governor Cabello wrote to Captain Cazorla that he had learned that the sergeants who had monitored the roundup had lied; over 1,300 head had been gathered, many of which were from private ranches and others were orejanos. Cabello then ordered testimony from several individuals who had helped gather the cattle and drive them to San Juan Bautista del Río Grande. Their statements indicated that 1,314 head of mature cattle and 294 yearling bullocks and heifers had been gathered. Two-thirds of the mature cattle and all the yearlings were orejanos. Of the branded cattle, some were from the mission, and others were from private ranches. In December 1784 Cabello officially accused Ecay of fraud and Cazorla of complicity. By July 1785 officials in Mexico City had concurred. However, it appears that Ecay found reasons to travel, and in September 1786 Cabello ordered that a request be sent to the governor of Coahuila to have Ecay appear in San Antonio for trial as soon as he returned to Saltillo. It is unclear, however, that Ecay ever appeared for trial because in December 1786, over two years after the alleged crime, Governor Cabello turned the proceedings over to his successor.[23]

The San Antonio clergy were frustrated with their inability to control and multiply their mission herds. In 1785 Fray José Franco López complained to the Bishop of Nuevo León that both Indians and Spaniards took their toll on mission herds, though it was "not easy to ascertain who has eaten or killed off the most." Fray López placed blame on several groups: the Apaches, who killed "at least twenty a day"; Spanish carneadores (meat hunters), who killed "more than a hundred head, and sometimes two hundred" each week; those who "from month to month" sold "to the presidio more than a hundred and fifty beeves"; the soldiers in charge of the presidio horse herd, "who are not satisfied with two beeves a day for twenty men, but on occasions kill four"; the troops of the presidio, who caught unbranded calves for their own use; and private ranchers, "who have taken away [exported] whole herds during the last eight years, totaling more than fifteen thousand head, most of them cows."[24]

During the 1790s the government regularly issued permits to round up cattle on the ranges along the Río San Antonio. The number of permits increased from seventeen in 1791 to a high of forty-eight in 1793, then declined to eleven or fewer per year for the remainder of the decade. The Delgado, Arocha, Menchaca, Flores, and Guerra families, all local residents, applied for most of the permits. In 1793 Governor Muñoz, concerned that the number of orejanos was declining, requested a report. He was told by local citizens that compared with the numbers in earlier years, very few orejanos could be found—so few that they now had to be roped individually rather than be driven in herds into corrals. Despite this testimony, Muñoz complained that, though the number of ranches had increased, no more than ten ranchers in the area were maintaining tame herds of branded cattle. The rest were content to capture and kill or export orejanos. In 1795 the government, in response to lawsuits by San Antonio ranchers, again upheld Croix's 1778 decree, and Governor Muñoz confirmed that in one year all unbranded cattle would become property of the Crown. Recognizing that this would encourage the slaughter of unbranded cattle for meat and tallow, he ordered that all sales of tallow be reported to authorities, and anyone reporting illegal slaughters would receive the meat, tallow, and half the fine.[25]

Despite reports that herds were being depleted, prices of cattle did not rise. Between about 1775 and 1800, the prices of small numbers of cattle sold to the presidio to feed visiting Indians hardly fluctuated, ranging from four to five pesos per head. This was considerably higher than the two to four pesos per head paid for herds sold in Saltillo or in the San Antonio area during the same period, and it may have resulted from the close ties between the ranchers and the military. Almost certainly, the availability of cattle from ranches between the Río Grande and Río Nueces kept prices from rising in spite of declining cattle numbers near San Antonio.[26]

The disputes over ownership and taxes owed for slaughtering, branding, and exporting cattle were symptomatic of the inherent conflicts among missionaries, civilian settlers, and the military in Texas. In fact, the settlements along the Río San Antonio were built on conflict. The Spanish government opposed France by establishing missions and presidios and prohibiting trade with Louisiana. Missionaries struggled to attract and hold often-reluctant Indian neophytes. Isleños conspired against non-isleños to maintain control of villa government and irrigated farmland. Farmers were in constant conflict with ranchers over damages to crops and livestock. Missionaries struggled with private ranchers over ownership of unbranded cattle. The government and missionaries continually pressured vecinos to pay taxes and tithes. Finally, the entire community lived under threat of Indians, who took advantage of every opportunity to steal livestock and sometimes killed or wounded Indian neophytes, vecinos, and soldiers during the raids. These conflicts weakened the community and contributed to its subsequent decline. However, during the last half of the eighteenth century a somewhat different model of ranching community was developing along the Río Grande and northward to the Río Nueces.

Chapter 4

SETTLEMENTS BETWEEN THE RÍO GRANDE AND RÍO NUECES

As ranches were developing along the Río San Antonio, another ranching frontier was growing to the south along the Río Grande. As early as the 1730s the ranching frontier had begun to move northward from Coahuila and Nuevo León into the lands along the Río Grande. As a result, in 1746 the province of Nuevo Santander was organized, reaching northward from Tampico across the Río Grande to the Río Nueces. Hoping to populate the region at the least possible expense, the government awarded a wealthy Spanish nobleman, José de Escandón, the right to settle the province. In January 1747 seven companies of soldiers, each led by a captain selected by Escandón, left different towns in northern Mexico and Texas and converged on the valley of the Río Grande, where they evaluated possible sites for settlements. In October, Escandón recommended that the region be colonized with soldiers who had participated in the reconnaissance. The report was well received by the government, and in 1749 Escandón led a large expedition of about 2,500 settlers and 750 soldiers to settle the region. Camargo, at the confluence of the Río San Juan and Río Grande, was the first town to be settled, under the leadership of Captain Blas María de la Garza Falcón. Down the river, Reynosa was established under Captain Carlos Cantú. Within six months, thirteen towns, each with a mission, had been settled between the Río Grande and the Río Panuco.[1]

By the time Escandón returned in 1750, Camargo and Reynosa together had more than 100 families and 36,000 cattle. By 1755, he had founded twenty-three settlements with 1,481 families and a population of 6,383 in Nuevo Santander. In addition, there were 2,837 Indians living in fourteen missions. Two years later the province was home to 8,993 settlers, who owned 58,000 horses, 25,000 cattle, and 288,000 sheep and goats. The largest sheep and goat ranches were along the Río Grande, and ranching soon began to expand northward from the river.[2]

Escandón's approach of awarding large land grants to his captains was, of course, in marked contrast to using missions and presidios as the vanguards of settlement. By giving the captains substantial control over the defense, economic development, and religious affairs of the settlement, Escandón avoided much of the conflict that continued to plague San Antonio and La Bahía. Initially, neither the captains nor the settlers held title to the lands they occupied. However, in 1757, at Escandón's recommendation, a commission led by José Tienda de Cuervo was sent to the region to obtain information and make recommendations for future government actions, including awarding land grants. The commission found that agriculture had been a failure at Reynosa because irrigation could not be developed. Similar problems were observed at Camargo, though some crops had been harvested in the river bottom. However, ranching was flourishing. Seventeen ranches had already been established in the vicinity of Camargo. Its mission now included 200 Indian neophytes from five tribes, and it even boasted a soap factory. The land around Mier was not suitable for agriculture, but the settlers had developed ranches and traded hides and livestock to the Indians in exchange for salt. Settlers at Revilla had established twenty-nine ranches and were growing all the food they needed. On the north side of the river twenty-three families at Dolores and fifteen at Laredo were engaged in ranching.[3]

Despite strong recommendations by Escandón and the Cuervo commission, land titles were not granted by the government until, at Escandón's insistence, a second commission visited the region in 1767. The first land grants were in the form of long rectangular parcels called *porciones,* with the narrow end bounded by the river to allow each settler to have access to its water. Porciones were laid out on each side of the towns and on both sides of the river. Settlers from Camargo were allotted 111 porciones with an average width along the river of 1,500 varas, stretching 12,500 to 20,000 varas away from the river (20,000 varas equals about 10.4 miles). Reynosa settlers received 80 porciones with an average width of 1,250 varas and depths of 20,000 varas to 25,000 varas. The sizes of these first grants were determined by the length of time the grantee had lived in the area, with the captains and original settlers who had come with Escandón receiving larger grants than more recent arrivals. A total of more than 300 porciones was awarded in the valley of the Río Grande, with an average area of about 6,400 acres.[4]

After the first porciones along the river were awarded and adjudicated in 1767–68, prominent individuals began to request and receive much larger land grants to pasture their growing herds. These were mostly located to the north of the original porciones in the brushlands between the Río Grande and the Río Nueces. Captain Juan José Hinojosa, his son-in-law Captain José María Ballí, and their descendants had the relationships and the financial means to successfully petition for large grants, and they were soon among the most prominent south Texas families. Some of the larger grants included La Feria (1777, 12.5 leagues), Espíritu Santo (1781, 59.5 leagues), Concepción de Carricitos (1789, 11.5 leagues), Llano Grande (1790, 25 leagues), San Juan de Carricitos (1792, 136 leagues), San Salvador de Tule (1797, 72 leagues), Las Mesteñas Petitas y la Abra (1798, 35 leagues), and Santa Anita (1800, 21.5 leagues). In addition, a number of smaller grants were awarded from about 1770 to 1810. Jackson has documented more than 130 ranches that were founded between the Río Grande and Río Nueces prior to about 1810.[5]

Some south Texas *rancheros* accumulated large herds of sheep, goats, horses, mules, and cattle. In 1757 Vásquez Borrego's Tienda de Cuervo hacienda at Dolores had 3,000 cattle, 3,000 mares, 400 stallions, and 1,600 mules, all cared for by 23 families. By 1767 the hacienda was producing 500 to 700 mules per year. By 1762 Nicolás de los Santos Coy, one of the original settlers at Camargo, had accumulated about 3,000 sheep, 2,500 goats, 400 horses, 300 mules, and 140 cattle. José Narciso Cavazos received the huge San Juan de Carricitos grant in 1792. By 1807 he owned 6,400 sheep, eight manadas (each composed of about 20 mares and a stallion), 13 male and 4 female donkeys, about 5,000 unbranded cattle, more than 50 mules, and unknown numbers of sheep. Despite having numerous livestock, some ranches had minimal improvements, only enough to sustain the families that cared for them. For example, José María Ballí owned Rancho de la Soledad de la Feria in present-day Hidalgo County. When he died in 1788, his will listed fifty manadas of mares, another 100 horses that were probably used by vaqueros, almost 50 donkeys, 200 branded cattle, and 2,000 sheep that were leased. Despite the numerous livestock, the ranch's improvements were listed as only three jacales, four corrals, and a fenced, irrigated field.[6]

An inventory of Hacienda Santa Anita taken when its patriarch, Don Manuel Gómez, died in 1803 (before the title was formally confirmed) gives us an idea of the relative values of land, livestock, and other articles of daily life on a south Texas ranch that specialized in raising sheep and mules. It is noteworthy that livestock made up over 80 percent of the property value.[7]

- fifteen sitios (de ganado mayor) and six "porciones" of land (284 pesos)
- buildings, including a demolished house (134 pesos)
- a cannon, pistols, saddles, bridles, copper kettles, a steel pot, an iron griddle, religious items, tools (198 pesos)
- livestock—including horses, burros, mules, sheep, goats, and cattle (3,209 pesos)

AGRICULTURE ALONG THE RÍO GRANDE

In the late 1820s, an unsettled period following Mexico's independence, Laredo had a population of just over 2,000. Jean Louis Berlandier reported that annual food consumption was about 365 cows; 100 head of sheep, goats, and pigs; 700 arrobas of wheat flour; and 6,500 fanegas of corn. The livestock were obtained locally; the municipality contained over 2,500 cattle, 2,000 sheep and goats, 200 hogs, and 500 horses and mules. But the flour came from Coahuila, and some of the corn had to be brought from San Antonio because irrigation was impossible and local production was only 4,000 to 5,000 fanegas. The Río Grande frequently flooded, changing course within its bed, which was "at least two hundred varas wide," and destroying any irrigation dams or canals that had been built. Because farmers could not plant their crops "on the dry and arid plains where corn does not grow," they planted in the moist fertile soils of old channels cut off from the flow of the river but "refreshed by the waters." Unfortunately, "every year or at least almost every year" floods destroyed the fields. Though farmers sometimes attempted to grow a second crop, summer rains would usually either "flood the country" or be "so scarce" that they were "of no use whatever for agriculture."[8]

Near Laredo, even cattle raising was difficult. Unlike areas farther north and east, the grazing lands were "covered with grasses" only in the spring. During the rest of the year the cattle were forced to eat cactus pads, or *nopales*. Berlandier observed that when there was "nothing but old, very prickly nopales, the herders gather them in large piles and singe off the thorns by holding them over a fire." The singed pads, "full of sap," were "excellent fodder, especially for cattle."[9]

In 1829 Berlandier visited the towns along the Río Grande. He noted that in 1828 Revilla had 3,167 inhabitants, and the 44 ranches in the municipality had "extremely numerous" herds, including "2,056 horses, 809 pigs, 586 mules, 1,349 head of large livestock [probably referring to cattle], and 32,301 head of small livestock [sheep and goats]." In addition, an average of "forty fanegas of corn and six of beans" was planted each year. Mier had 2,831 inhabitants in 1828, mostly farmers and herders. Their crops, like those in Laredo, were grown without irrigation near the river "where floods sometimes carry off or submerge their harvests." Camargo, down the river from Mier, had 2,589 inhabitants and "several hundred thousand animals of all kinds" before Mexican independence, and "numerous huts, surrounded by fields and herds" dotted the lands, primarily north of the river. However, its population had decreased and its herds had been reduced to "scarcely twenty-five thousand head" since "the cruel and lengthy war of the Comanches and the Lipans towards the last years of Spanish domination." Thirteen or fourteen leagues down the river was Reynosa, whose fields were also plagued by floods. Twenty-five leagues farther down the river Matamoros, formerly known as Congregación del Refugio, had grown to a population of perhaps 6,700 since Mexican independence, in large part due to its growing importance as a port. Located near the coast, the town suffered numerous diseases. Ships brought merchandise, primarily from the United States, but "the industry of the inhabitants of that town and of its surroundings" was "practically nil," though they had begun to sell some hides for shipment to the United States.[10]

Even though Berlandier reported no significant irrigated agriculture along the Río Grande, by the 1840s it had begun to develop. Cotton was becoming an important crop near Matamoros, and "sugar and cotton plantations" were reported by visitors near that city. Miguel Salinas took legal action against the U.S. government, claiming that during the Mexican War U.S. troops stole property from his hacienda, part of the old Espíritu Santo grant. In his deposition he claimed to have four thousand fruit trees and to produce sugarcane, corn, beans, and vegetables. He also owned a sugar mill and stated that his annual production was worth twenty thousand dollars.[11]

Despite some agricultural development along the Río Grande, Teresa Vielé, traveling up the river in 1852, described a rancho near the river above Edinburg as "half a dozen mud-huts neatly thatched with straw and open sheds attached for culinary purposes, where the kettle hung suspended over a wood fire." The women milked goats and cows, and "a large garden and a good-sized patch of Indian corn, interspersed with melon vines, together with cattle and an enormous flock of barn-yard fowls, completed the scene." The peasants built their jacales "of straw, reeds, stone and adobes, without either nails or hammer." They practiced simple subsistence agriculture using as a plow "a sharp-pointed log, with a pole at one end" to guide it and another pole at the other end, to which a pair of oxen was "strapped by the horns." Their harrows were "made of the branch of a tree," and their corn was "put into the ground and left to Providence to either ripen or dry up, of which there [was] an even chance."[12]

FROM MATAMOROS TO THE RÍO NUECES

After Mexico gained its independence in 1821, many of the ranching families who had fled south of the border during revolutionary times wished to return. In 1824 the region from the Río Grande to the Río Nueces became part of the Mexican state of Tamaulipas. But the Apaches and Comanches had taken advantage of the unsettled political situation in Mexico. Berlandier reported that between the Río Grande and Río Nueces in the early 1830s "few dwellings in these deserted areas [had] not been watered in the blood of some traveller or colonist." "Before the cruel and lengthy war of the Comanches and the Lipans [Apaches] toward the last years of Spanish domination in these regions, numerous huts, surrounded by fields and herds, dotted the now deserted lands," but now the ranches north of the Río Grande were "almost deserted." Only "some miserable huts (ranchos), [were] scattered along great distances and inhabited by herders who tend their horses or cattle." Whereas "in other times there were several hundred thousand animals of all kinds in the jurisdiction of Camargo," mostly north of the Río Grande, by the 1830s "all the herds together scarcely number[ed] twenty-five thousand head." Only cattle, goats, and sheep were raised between the Río Grande and Río Nueces because domesticated horses and mules were either taken by the Indians or were led "off into the wilderness" by herds of wild horses. On the road between Matamoros and the Río Nueces, wild horses, or mesteños, were "encountered in greater abundance." One herd was composed of "at least one thousand to twelve hundred horses of all ages, as well as some runaway mules that lived with them." But as the ranches were reoccupied, their herds increased. Horses and mules from ranches in Tamaulipas began to be exported from Matamoros to New Orleans, St. Louis, and other parts of the United States. In 1835 breeding mares sold for two to five dollars apiece. Good saddle horses brought eight to ten dollars, working oxen could be purchased for fifteen to twenty dollars a pair, and sheep sold for fifty cents apiece.[13]

However, the Indians still threatened isolated settlements. For safety, ranches were often grouped together. For example, Rancho Los Ojuelos (Little Eyes or Springs), was located in present-day Webb and Jim Hogg counties northeast of Laredo. Based on a two-league grant to Eugenio Gutiérrez in 1810, by the 1850s Los Ojuelos had more than doubled in size, to over twenty thousand acres. Its owners, the Guerra family, had established their headquarters at the base of a rocky ridge, or escarpment, with a number of small springs, or *ojos de agua* (eyes of water). The family had constructed two rows of stone houses to protect their springs and themselves from hostile Indians, and over four hundred people, including the Guerra family and their workers, lived on the ranch. Other ranches were established nearby, both to protect the families and to take advantage of other small springs in the area. Several of these had names indicative of the water so necessary for life in this dry region. They included Las Albercas de Arriba (Upper Pools), Las Albercas de San Felipe (San Felipe Pools), El Bordo (The Edge, referring to the ridge), El Pato (The Duck), El Patito (The Little Duck), and La Tinaja (The Pond).[14]

Another grouping of ranches was established in present-day Jim Hogg County after 1830 by residents of Mier. Rancho El Randado, established in 1836 by Hipólito García, initially covered forty-five thousand acres, but it grew, eventually occupying over one hundred thousand acres at García's death in 1888, one of the largest and most prosperous ranches in Texas. The Benavides family's Norecitas and Sigifredo Muñoz's Las Cuevitas were other ranches in the area. Residents of Mier established another group of ranches in present-day Duval County. Julián and Ventura Flores were granted a total of eight leagues of land in 1812. They founded two ranches, San Diego de Arriba and San Diego de Abajo, separated by San Diego Creek. By 1844 the community had about twenty-five families, and in 1848 Pablo Pérez bought the land on the north side of the creek. Pérez soon married Vicente Barrera, inheriting Rancho Amargosa and established two more ranches, Los Reales and El Muertecito. Other ranches in the area were Las Conchas, La Trinidad, Santa Gertrudis, Petronilla, Concepción, La Rosita, Mendieta, Peñitas, Veleño, and Lagarto. These ranches had thousands of cattle, horses, and sheep, and by 1850 San Diego was an important trading community, where the roads from San Patricio to Brownsville and from Corpus Christi to Laredo crossed. Residents of Reynosa, families such as the Hinojosas, Ballís, Garzas, and Canos obtained huge Spanish land grants, intermarried, and developed ranching empires. The ranches developed by these families eventually occupied lands along the Gulf Coast all the way to the Río Nueces, including Padre Island, named for its owner Padre Nicolás Ballí.[15]

LIFE ON THE RANCHOS

Texas historians have long emphasized the debt Texas ranching owes to Hispanic ranching practices of the Mexican plateau. Franciscan missionaries had ties to Querétaro and Zacatecas, and most of San Antonio's trading ties to the south were with Saltillo and other cities of the plateau. The sheep- and goat-raising traditions of the plateau found a home around Laredo and the other inland towns and ranches along the Río Grande, where sheep greatly outnumbered cattle. The flat-roofed, fortified ranch houses from the plateau were well suited for the drier western areas of Texas. However, historian Terry Jordan has pointed out that the ranching traditions along the Gulf Coast of Mexico greatly influenced the cattle-raising industry between the Río Grande and Río Nueces. Many of the families recruited by Escandón were from the cattle-ranching areas along the Mexican Gulf Coast north of Tamaulipas. In both coastal areas cattle raising predominated, and many words associated with Mexican cattle ranching made their way, often with modification, into the Anglo-Texan vocabulary, such as *tanque* (tank), *lazo* (lasso), *la reata* (lariat), *mesteño* (mustang), *mata* (motte), and corral. In addition, Jordan points out that cattle herds from the coastal areas south of the Río Grande were the source of many Texas cattle.[16]

Typical ranch families in south Texas during the last half of the eighteenth century raised cattle, horses, sheep, goats, burros, and mules. They grew much of their food, including corn, beans, squash, chilies, and spices. Some ranches, especially in drier inland areas, specialized in sheep and goats. Along the coast between the Río Grande and Río Nueces, where higher rainfall supported richer grasslands than farther inland, ranchers raised more cattle and mules. Usually ranches were highly interdependent, multigenerational operations, and the land was typically held in common by the family. But the ranch was usually guided by the oldest male in the family (*el patrón*) or, less commonly, the family matriarch (*la patrona*). On smaller ranches most of the labor was provided by members of the extended family. Some larger ranches were, however, true haciendas, with two distinct social classes, the family of the patrón and the *peón* class, composed of laborers, vaqueros, and pastores.[17]

The typical ranch headquarters consisted of a complex of houses, other buildings, and corrals. At the functional center of the complex was the *casa mayor,* or large house of the patriarch. Beside the casa mayor was the patio, the focal point for much of the work in the household. Often covered with flagstones, the patio usually contained a brush-covered arbor, or *ramada,* to provide shade during the heat of the day. Across the patio from the house was a stone chimney for food preparation and an oven, or *horno,* to bake bread and burn seashells for lime. A *corral de leña,* or log corral, was often near the patio. This strong, durable structure was built by setting pairs of vertical mesquite posts five to eight feet tall and about three to six feet apart. Smaller mesquite logs were placed horizontally between the vertical posts, forming a solid log fence capable of holding most animals. Around the casa mayor were smaller houses, usually simple thatched jacales of the other family members. A large vegetable garden was located near the casa mayor and provided a variety of vegetables and herbs. Corn, beans, squash, chilies, onions, garlic, cilantro, and watermelons were complemented with a number of medicinal herbs. Smaller gardens might be located near the jacales. The ranch yard was often enclosed by a strong stockade or fence to keep out wild animals. Outside the yard were corrals, wells, and sometimes a chapel, commissary store, school, and family cemetery.[18]

In most cases the casa mayor was a *casa de sillar,* a house with walls made of large, rectangular, cut limestone blocks measuring about two feet long, twelve inches wide, and eighteen inches high. These stones were excavated from the horizontal beds of limestone that underlie

Casa de Sillar

much of south Texas. They were relatively soft when excavated and could be cut with saws, hammers, and chisels. They were then moved by oxcart to the homesite. The typical casa de sillar was at least sixteen feet by twelve feet with one door and two windows. The floor was usually made of large flagstones, and the flat roof was of *chipichil,* a cement composed of sand, lime, and gravel. This cement roof was supported by heavy beams, or *vigas,* made of mesquite or cypress covered by hand-hewn boards called *tablas.* The chipichil roof was very durable and sloped slightly to drain rainfall through downspouts into the cistern. For defense against Indian raids, gun ports, called *troneras,* were placed in the ends of the house and beside the doors and windows to allow occupants to fire on attackers. The walls of the house normally extended three to five feet above the flat roof to provide protection for residents defending the house from there.[19]

Peónes, as well as members of the patrón's family who did not live in the casa mayor, usually occupied one- or two-room jacales. Each room was eight to twelve feet on a side. The structure was supported by four corner posts buried solidly in the ground and forked at the top to support heavy vigas, or roof beams atop the walls. The walls were made of whatever materials were available, often small mesquite logs set vertically in the ground. Flexible branches were sometimes woven horizontally between the vertical posts to provide additional strength. The wooden walls were normally sealed with a mixture of clay and lime, often mixed with grasses for strength, covered with a protective coating of cement made of lime and sand, and whitewashed. The floors were usually packed earth. Two tall posts set in the ground supported the ridgepole of the gabled roof. Rafters were tied from the ridgepole to the vigas, and light poles were tied horizontally across the rafters to support the grass thatch. The typically rudimentary furnishings consisted of a handmade bed with grass mattress, a table and chairs, a mirror, and the family altar.[20]

Because securing water for the inhabitants and their livestock could be difficult, ranch headquarters were often established near springs. When springs were not available, *norias* (wells) were dug. These were sunk to the water table, and the first few feet, down to the level of solid limestone, were lined with sillares. For small household wells, water was drawn by hand with a wooden bucket, or *cubeta.* For larger livestock wells, a rawhide tub called a *buque* was lifted by a rope tied to the saddle of a horse or mule. When it reached the top of the well, the buque was tipped, and the water flowed into a long, stone water trough. When water was scarce, vaqueros worked in shifts to draw enough water for the livestock. Another method of obtaining water for livestock was to

Jacal

build a dam, or *presa,* across a dry streambed. Fill to make the dam was dug with hand tools and moved in a *mecapal,* a rawhide or burlap bag container carried on a worker's back and supported by a leather forehead strap. A larger rawhide container, the *guaripa,* could be carried by four workers or could be dragged between two horses. A typical presa could take months of labor to complete, but the resulting reservoir, typically up to twelve feet deep and sometimes covering several acres, would capture runoff from the region's infrequent storms and store it for months.[21]

Food in Hispanic south Texas was almost entirely dependent on the crops and livestock from the region. The three principal foods were corn tortillas, dried meat, and beans. To prepare tortillas, corn grains were boiled in lime water made by pouring water through wood ashes. This softened the grains and allowed their tough, yellow skins to be removed. The remaining material, called *nixtamal,* was ground between a hand stone, or *metapal,* about nine inches long and three inches in diameter and a rough stone *metate* about eighteen inches long and twelve inches wide. The nixtamal was then patted into flat, round tortillas and cooked on a *laja* (flat stone) or *comal* (cast-iron griddle) heated by coals. These tortillas could be eaten alone, folded and used as a spoon to scoop up other foods, and used in preparation of other dishes. Noah Smithwick described a Mexican woman "down on her knees before a little bed of glowing coals on which lay a piece of sheet iron on which a couple of tortillas were baking." The woman had a bowl of hulled corn and a metate, "simply two flat stones between which the softened corn [was] mashed into dough." The woman produced the tortillas "with the regularity of clock work, taking off one, turning another and putting on a third, then preparing another" at the metate.[22]

Dried meat (*carne seca* or *acecina*) was prepared by cutting very thin slices of meat, salting the strips, and drying them for several days by hanging them across a rope in the sun, taking care to separate the two sides of the strips with a small stick. The dried strips could be tenderized by pounding and would not spoil for weeks if kept dry. Brown beans (*frijoles*) were cooked for hours in a cast-iron kettle, which could usually be found simmering on a fire in the kitchen. Spices were ground by crushing them in a shallow stone *molcajete* about eight inches in diameter with a hand-held pestle (*mano*). Other basic kitchen equipment included a butcher knife and coffeepot.

Typical daily meals included two breakfasts (*desayunos*), a light breakfast of bread and coffee early in the morning and a heavier breakfast at midmorning; a light meal at midday (*almuerzo*); and a dinner (*cena*) in the evening. Carne seca could be eaten in several ways: rolled in a tortilla as a *taquito,* usually while a person was working; shredded and mixed with scrambled eggs (*machacado con huevos*); and cooked in spicy stews containing chilies and squash. A variety of other meats were available on ranches, including cabrito, mutton, wild game, and beef tongue, head, brains, tripe, and stomach. Other favorite foods included *tamales* made of nixtamal placed on a corn shuck, then wrapped around the meat from beef heads, and baked overnight in a deep pit. *Enchiladas* were made by wrapping tortillas around shredded meat or cheese and then baking. *Menudo* was a favorite soup made of beef stomach and hominy. Vegetables included squashes (*calabazas*), chilies (*chiles*), and sliced strips of cactus pads (*nopalitos*). Sweets commonly consumed were ripe cactus fruits (*tunas*), piloncillo, and flour tortillas that were fried and dusted with sugar and cinnamon (*buñuelos*).

Although most of the foods consumed in south Texas were produced locally, a few had to be imported from the south, including wheat flour, coffee, cocoa, wine, distilled spirits (aguardiente), and spices such as cinnamon. In addition, arrieros sometimes brought special dried chilies and piloncillos, for which local production was insufficient.[23]

THE VAQUERO'S TOOLS

The vaquero's saddle (*silla*) was perhaps his most important piece of equipment, having evolved over more than two hundred years from the Spanish war saddle (*estradiota*) of the conquistadores and the lighter

Moorish *jineta.* The war saddle was still in use in Texas in the late eighteenth century. Designed to protect the soldier, it had a tall rear cantle to keep the soldier from being pushed off the rear of the horse during battle. The front part of the saddle was also tall to protect the rider's lower torso. Stirrup leathers were long to allow the rider to straighten his legs and brace himself against the cantle. Pagés described the heavy war saddle used by soldiers in east Texas in 1767 as "neatly dressed and stamped with various ornamental designs . . . garnished round the edges with trinkets of steel, which like as many little bells, are kept perpetually ringing by the motion of the horse." The stirrups were "composed of four massy bars of iron arranged in the form of a cross" and weighed "at least fifty pounds" in order "to keep the horseman steady in his seat, and to constrain his limbs to that position which is deemed most graceful among the Spaniards." Riding in the Spanish style was a "pretty severe trial" to the novice and caused a "swelling of his legs, and an almost entire dislocation" of his joints. Nevertheless, Pagés felt that "if the horse is strong enough not to be oppressed by their weight," the heavy stirrups might "contribute to his [the horse's] ease, since they form a sort of balance below the gravity of the rider on his back."[24]

By the late eighteenth century, the saddles used by vaqueros on the northern frontier had become smaller and lighter than the war saddle to accommodate the vaquero's need for mobility on horseback. The

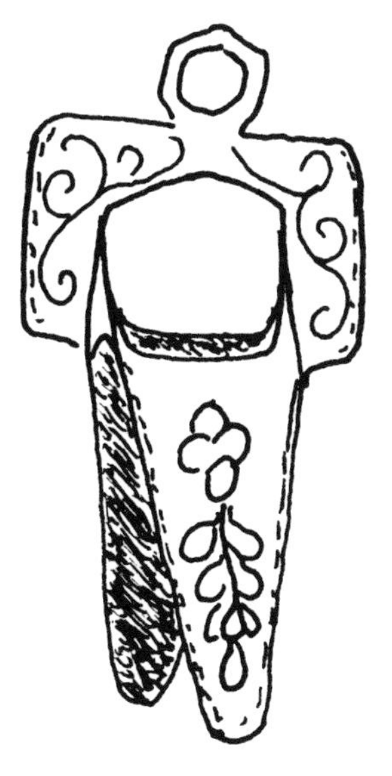

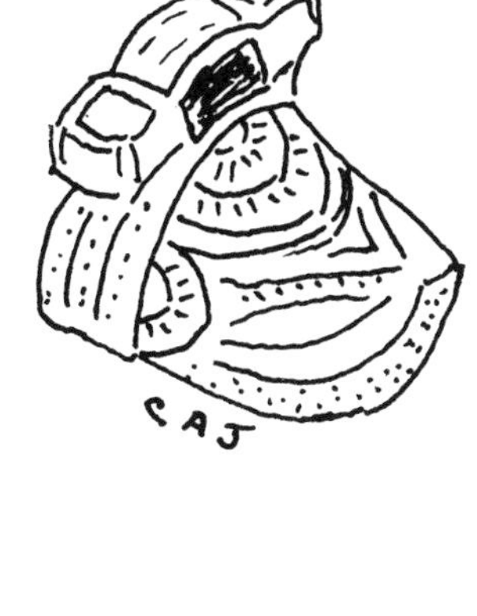

Stirrups

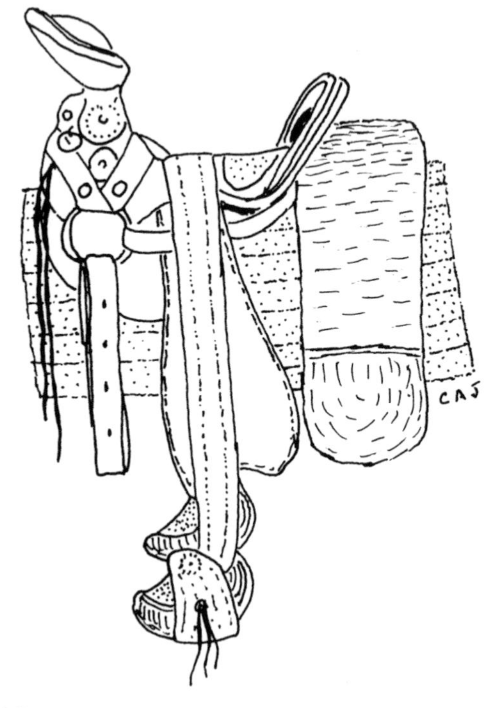

Saddle

wooden saddletree consisted of two sidepieces joined in the front by a wooden fork and in the rear by the cantle. These four pieces were formed into a single rigid unit by stretching and sewing wet rawhide over them. A pommel or horn (*poma*) at the front of the saddletree provided a post around which a rope could be secured. The saddletree was covered with leather pads and skirts to protect the horse's flanks and sides. The stirrup leathers were looped around the sidepieces of the saddletree and held heavy wooden stirrups that were closed in front and often carved in the shape of a lion's or other animal's head. Leather stirrup covers provided extra protection to the rider's feet. The horsehair cinch was attached by a leather strap to a ring that hung from two leather straps secured to the front and rear of the saddletree. This means of attaching the cinch to the saddle was later known as "Spanish rigging." The saddletree, rigging, stirrup leathers, pads, and skirts were covered with a removable leather cover, and an ornamental leather "duck's tail" was attached to the rear of the saddle, covering the horse's rump and protecting items tied behind the saddle from the horse's sweat. Small saddlebags were often attached to the pommel and hung in front of the rider's thighs. The rider tied pistols to the rear of the saddle and attached a short rifle (*carabina*) in its case to the front. In the nineteenth century the saddlebags and rifle would be moved behind the saddle. As protection from thorns, rawhide covers could also hang from the pommel and be tied over the rider's legs to the back of the saddle. These later evolved into the *chaparreras* (chaps), worn by the rider rather than attached to the saddle. In the early 1800s Zebulon Pike observed that the vaquero's saddle was superior to the saddle used by the English and vaqueros were "probably the best horsemen in the world."[25]

After the saddle, perhaps the most useful piece of equipment used by the vaquero was the reata, a braided rawhide rope later called a lariat by Anglo-American cowboys. Like the saddle, the reata evolved over the years. Originally the reata was a loop of rope placed over the animal's head with the aid of a garrocha, but vaqueros soon learned to make a loop in the end of a long rope and throw the lazo over the animal's head. At first the reata was tied to the horse's tail, but it soon became clear that cattle could be controlled more effectively when the reata was secured to the saddle horn. Vaqueros made their own reatas by cutting long, thin, continuous strips of rawhide beginning on the outside of the hide, shaving the hair off with a knife, wetting and stretching the strips, and then braiding them into a rope about the size of the little finger. As soon as an animal was roped, vaqueros would quickly give the reata a turn (*dar la vuelta*) around the poma. Dar la vuelta was later anglicized to *dally* by Anglo-American cowboys, and those who were not experts could easily lose a thumb caught between the rope and the saddle horn.[26]

Vaqueros also used ropes called *cabrestos* that were woven from horsehair. In the early 1820s Elias Wightman described the Texas vaqueros as "most admirable horsemen." Mustangs were caught by "throwing the cabaresto [cabresto] or hair rope . . . seldom missing at a distance of thirty yards." The vaquero preferred that the animal to be caught "be under full motion." As soon as the noose settled over the animal's head, the vaquero, "whose saddle is buckled as tight as can be with a strong hair girth and one end of the cabaresto around the saddle horn, falls back a little and passes on the other side of the mustang, dashes forward drawing hard on the cabaresto, which throws the mustang." A second vaquero would then "lash and tie the creature thus caught" before it could rise. Wightman reported that for amusement, the vaqueros would "take after any of their own stock, which are almost perfectly wild . . . catch a cow, steer or bull by the tail, fetch it a twist around their saddle horn, dashing past along side and throwing them heels over head and give chase to a new one."[27]

During the period of relative peace after 1785, vaqueros were able to hunt the wild horses common on the Texas prairies. Visiting Texas in 1806, Pike reported large herds of wild horses, or *mesteños,* between the Red and Colorado rivers and described the methods used by the vaqueros to capture them. They took "a few fleet horses . . . into the country where they expect to find the wild ones." There they built "a large enclosure" with wings projecting outward from the entrance "a great distance" onto the prairie. Bushes were cut and placed strategically "to induce the horses when pursued to enter these wings." After the pen and other preparations were complete, the vaqueros kept "a look out for a small drove, say two or three hundred." It was important not to try and capture a larger herd of several thousand head because they would "either burst open the pen, or by crowding, pressing and running over one another" would kill the weakest animals, and "the pen will be filled up with their dead bodies and the survivors will run over them and escape." In addition, the "pen besides would be rendered useless for a long time afterwards by the stench." If, however, the vaqueros succeeded in "driving in a small drove of some few hundred[,] they select the youngest and handsomest, noose them and take them into the small enclosure, where by starving them, preventing them from sleep, [and] keeping them in continual motion, they are rendered gentle by degrees, and finally induced to submit to the saddle and bridle."[28]

Berlandier reported in the 1830s that these *mesteñeros* (wild horse catchers) led "a vagabond and lazy life," spending five or six months on the plains "nourishing themselves with the results of their hunting or even with the flesh of young colts." The operation described by Pike and Berlandier was very dangerous, both to the horses and the mesteñeros, and many horses were often killed. Berlandier reported that it was "rare that in one of these chases a large part of the horses thus trapped do not kill one another in their efforts to escape." He recalled one corral near the Río Nueces was "strewn with the bones of horses" that died there.[29]

In the 1830s Charles Sealsfield gave graphic descriptions of how Texas vaqueros broke the horses they had captured. First the horse had to be rendered tame enough to be caught. The process began when the vaquero roped the horse with "a strong noose or cord made of twisted strips of rawhide, and called a lazo." As soon as the loop of the lazo was throw over the horse's head, the loop was "instantly drawn tight, and the poor creature [was] thrown violently down without the power of moving, and generally deprived of its breath." Horses were "sometimes badly injured, and even killed by being dashed to the ground; but generally escape[d] with a severe practical lesson on the nature of this rude instrument of civilization." A horse that had been roped and choked down by vaqueros never forgot the experience. The mere sight of a lazo would make the horse "tremble in every limb," and simply showing it the rope or laying it on the horse's neck would "render him as tame and docile as a lamb." The next step was breaking the horse to ride. After it was roped, the horse was blindfolded to prevent it from rearing or trying to run. It was then saddled and bridled. The vaquero then mounted the horse and pulled off the blindfold. To control the horse, vaqueros used bridles with "terrible lever bits" that with "one pull, and not a very hard one either" could tear its mouth "to shreds and cause the blood to flow in streams." Using spurs with long, sharp rowels, the vaquero spurred the horse and forced it to run until exhausted. After a few minutes' rest, it was mounted again, spurred, and forced to run back to where it began. Sealsfield reported that "if he breaks down during this rude trial, he is either knocked on the head or driven away as useless; but if he holds out, he is marked with a hot iron and left to graze on the prairie." A horse broken in this way remembered the experience, and "there was no particular difficulty in catching him" afterward.[30]

Another visitor from the United States wrote that "no people are equal to the Mexicans in the management of their animals when wild and apparently ungovernable." He told of an American's mule "that had thrown every person who had ever attempted to ride it . . . a powerful animal with much more cunning than is common to its kind and malicious in the extreme." When mounted by "a large muscular African who conceived that nothing was an overmatch for him," the mule reared "upon its hind feet and then, bounding forward, threw its head down and its heels into the air, until it appeared to stand nearly perpendicular

Spur

upon the earth." Before the rider was "conscious of anything, he lost his balance and lay sprawling upon the ground." When it was clear that no one could ride the mule, a Mexican vaquero "stepped forward and prepared to ride it. . . . The moment the Mexican seized the bridle, the mule appeared to feel, instinctively, that he had at last found his master and cowered under his steady look, as he grasped it by the ear and gazed strongly upon it an instant before he mounted." As the mule began to buck, the vaquero "kept his balance with the most perfect composure." He pulled the mule's head to the side "until it touched its left shoulder and then, putting a heavy spur into its right flank, he whirled it round and round in a complete circle." After several spins, the vaquero reversed the mule's direction "until the mule, convinced that there was no alternative but go round forever or go quietly ahead, was satisfied to comply with the latter."[31]

The Tejano settlements and ranches between the Río Grande and Río Nueces by the early 1800s had eclipsed those along the Río San Antonio. The use of private ranchos, rather than missions and presidios, avoided many of the conflicts that limited development along the Río San Antonio. Many of the ranching practices developed or perfected in south Texas were later adopted by Texian cowboys and ranch owners, often on the lands and with livestock originally owned by Tejano rancheros.

Chapter 5

CONFLICT AND DECLINE

THE SAN ANTONIO MISSIONS REACHED THEIR maximum size and vigor in the 1760s, then began a long and often painful decline until they were secularized in the 1790s. The reasons for the decline included conflicts with private ranchers, the Apaches, and the Comanches and their Norteño allies; and epidemics of European diseases, including smallpox, measles, cholera, and influenza. Some Indians in the missions, longing for their old ways, fled the missions. Others became acculturated, married soldiers or villa residents, and moved away from the missions. In addition to dealing with the decline of the missions, military and civilian officials in San Antonio struggled with a variety of conflicts and trade restrictions that limited development of the community.

INDIANS

The Apaches and Norteños (Comanches, with their Wichita and Caddo allies) regularly harassed San Antonio and the surrounding area during the 1760s, 1770s, and 1780s. They constituted a serious threat and substantial drain on the farming and ranching economy of San Antonio and other communities in Texas and northern Mexico. This required that the able-bodied men of the villa assist in community defense and join parties pursuing bands of Indians that fled with stolen livestock. Although many of the men had their own arms, the Crown furnished muskets, swords, lances, and horses to civilians that could not afford their own.[1]

A few examples will illustrate how San Antonio, caught between the Apaches to the south and the Comanches and their allies to the north, suffered from and reacted to this hostile and ever-changing situation. In 1767 Pierre Pagés observed that the presidio was "but weakly fortified, and has much occasion for a stronger garrison." The Comanches "incensed against the governor on account of his prohibition of their trade with the French . . . carried off four hundred horses from San Antonio." To cover their tracks, the Indians "set fire to the sward . . . leaving three or four leagues of black desert behind them." On this occasion, troops pursued the attackers for about a hundred leagues without success, but on their trip home they were attacked by "another party of the same nation." The troops resisted the attack for about three hours but "at last yielded to superior numbers, and lost on this occasion, besides other property, a hundred and fifty horses."[2]

In 1770 fifty soldiers were transferred from Los Adaes to San Antonio "for defense against the enemies who commit hostilities on that land." However, a request for ten thousand pesos to strengthen the presidio palisade and add fourteen cannons was denied. In an attempt to protect the widely scattered ranches around San Antonio from the continued harassment by the Comanches, in the spring of 1771 Governor Ripperdá ordered the ranchers to select a common grazing ground for their herds and a common field in which to cultivate their crops.[3]

In 1773 the Comanches stole 100 horses from San Antonio and 272 from Laredo and then killed four residents of Laredo and Saltillo. Viceroy Bucareli soon became frustrated with the policy of "kind treatment and trade of arms and ammunition with the nations of the North." In May 1774 he ordered Ripperdá "not to be deceived by the knavish promises and feigned [overtures of] peace of the Indians," who had "stolen a number of horses from the . . . Presidio . . . so adroitly that to their satisfaction and pleasure, they were able to choose the best." Bucareli concluded that "with more than five hundred armed men whom Your Lordship can assemble," it should "be easy . . . to defend the province, and curb the pride of the Barbarians who commit [these] atrocities."[4]

For a better appreciation of the degree to which hostilities affected daily life, consider some of the events that disrupted life in 1775 and 1776. Two of six vecinos driving a herd of San Antonio cattle to supply distant presidios were killed by Indians. Soldiers and civil authorities narrowly prevented more than forty Apaches visiting Valero and Espada missions from attacking four Tawakonis and three Wichitas who arrived at the same time to visit the presidio. On the evening of April 14, 1776, a citizen was shot and killed by unknown assailants while retrieving a horse that was hobbled and grazing near the presidio. On April 23 the governor was informed that two settlers had been killed by Comanches in separate incidents near La Bahía and at El Fuerte del Cíbolo, halfway between San Antonio and La Bahía. On May 9 three soldiers from the presidio were attacked and one killed by Comanches. In mid-May five hundred Apaches raided the San Antonio and La Bahía ranches, killing large numbers of cattle. Continuing on to the lower Río Grande and up the river to Laredo, they stole horses and killed several people. Returning northward, they had the temerity to enter San Antonio to trade. The only thing the garrison could do was to try to prevent them from buying guns from the mission Indians. On August 4, 1776, an Indian from Mission San José was killed, probably by the same Apaches who were seen killing cattle nearby, a practice scorned by the Comanches. In mid-September three hundred Apaches passed up the valley on their way to hunt buffalo to the north, killing livestock and stealing corn from the fields as they went. When two hundred entered the villa to trade, all that the garrison could do was to prevent them from obtaining arms and ammunition. In October a detachment of soldiers encountered a Comanche camp about twenty leagues east of La Bahía, killing several Indians in two encounters.[5]

In October 1781 Croix reported that the "incessant raids of the Comanches are so terrible and bloody that if they continue at the same rate, the speedy ruin of the province" would soon result. He warned that "the whole province is overrun" by Indians; the ranches were "without stock" and were "rapidly being abandoned . . . [with] settlers taking refuge in the settlements."[6]

In May 1784 one hundred Norteños (probably Comanches, Wichitas, and Caddos) crossed the Río Guadalupe on their way to Espíritu Santo. There they encountered a group of twenty-five vaqueros gathering a herd of orejanos for export to Coahuila. Three times the Norteños threatened the vaqueros with "maneuvers that seemed to be nothing short of hostilities." The following evening the Spaniards, doubtless observing from a safe distance, reported that the Indians had set "eighteen fires . . . leaving a very suspicious smell [probably of roasting meat] as they devoured all the cattle that came within their reach." The vaqueros were "unspeakably apprehensive" and fled "into the wilderness" to avoid the larger Norteño force. At about the same time the vaqueros encountered a small party of Apaches, part of a larger force of about three hundred who were hunting south of the Guadalupe. When told of the Norteños, they "departed immediately to cut across the land" and inform their forces. On the way, they encountered five Norteños, killing four and capturing one. Learning that many more Norteños were in the vicinity, the three hundred Apaches "scattered in order to overtake their route." However, the Norteños, aware of the Apache presence, fled to the north. Frustrated, the Apaches "broke camp and moved to their old rancherias" between the Río Nueces and the Río Grande "with the intention of dancing and dining on" the captured Norteño.[7]

Another example of the Indians' audacity occurred in the summer of 1784. At dusk on July 8, fifteen Comanches and Norteños broke down a gate and entered the irrigated farmlands south of the presidio. When they found there "two poor and honorable residents . . . who were plowing, they killed them with spears and took parts of their scalps." Discovered by the residents, the Indians retreated to the northwest, pursued until nightfall by twenty-eight soldiers and residents. However, this did not end the troubles. At dawn on July 16 Indians broke a gate and entered the fenced garden of the house next to Governor Cabello's. Opening the gate to Cabello's garden, they made their way to his stable, where they took his "two best saddle horses and two others belonging to [his] servants." Cabello reported that "I stood for a while observing all this ruin, whereupon I was informed that [the Indians] had been in most of the gardens to the west, where they had wreaked indescribable havoc to watermelons, cantaloupes, squashes, and roasting ears, to such an extent that many of the owners were left unable to reap the most meager product of their diligent labor." To prevent further surprises, Cabello placed patrols of four men at each of the seven roads leading into the settlement. A few days later he dispatched forty troops, twenty-eight civilians, and fourteen Indians in pursuit, leaving the defense of San Antonio greatly weakened.[8]

Conflicts with the Indians decreased after 1785 when persistent Spanish efforts finally led to peace with the Comanches. From about 1790 through the early 1800s a lively trading relationship developed. Large numbers of Comanches came to San Antonio, camped on its outskirts, and traded. Hides, tallow, meat, captives, and horses were exchanged for textiles, clothing, ornaments, metal implements, weapons, and food. Services such as blacksmithing and gun repair could also be obtained in San Antonio. To encourage this trade, the Spanish authorities organized *ferias,* or trade days; constructed a large communal house for the Indians; and provided them with firewood, food, and tobacco. Spanish traders from San Antonio also ranged far into the Comanche homelands to exchange goods. The trade was so beneficial to all that Spanish authorities and Comanche leaders worked together to prevent clashes and punish isolated offenders. Peace with the Indians lasted until Mexico's struggle for independence began in 1810. Indians initially assisted the royalists because of their long friendship with the Spanish authorities; however, their loyalties later became divided as Spanish society disintegrated under the pressure of Mexican insurrectionists and American filibusters. As the war for independence came to a close, peace was reestablished, and by 1822 all the tribes had renewed their treaties with the Mexicans in San Antonio.[9]

CRIME

Indians were by no means the only source of violence in San Antonio. Despite the community's small size, San Antonio authorities had to contend with a wide variety of criminal activities, including vagrancy, drunkenness, gambling, robbery, smuggling, assault, and murder. To dissuade criminals, official decrees were periodically issued, setting out punishments for specific crimes. For example, in 1745 the alcalde ordered that anyone carrying arms "such as knives and the like for any reason or pretext whatever" would have to forfeit the weapon, pay a fine of twelve pesos, and be imprisoned for one week. However, should the offender "be dark colored like an Indian, mestizo, or the like," he would receive two hundred lashes and be forced to labor on public works for a month. Concerned that idleness was leading to mischief, he also ordered that "all bachelors, ruffians, and others living in this my jurisdiction without any work or source of income" had one week to find "masters whom they may serve" or suffer deportation. Finally, to protect against Indian attacks and other nocturnal transgressions, he ordered a curfew after eight in evening, explaining that "this is a region subject to war" and "if no one went out . . . after the customary time," it would be safer and easier to detect unfriendly Indians. Similar orders were issued in July 1749, and a decree in January 1754 noted that "there is hardly a house in this villa that is not used for gambling on the pretext of its being for amusement . . . and many are tempted to steal in order to gamble"; therefore, "there shall be no gambling of any sort."[10]

In 1770 Governor Ripperdá took note of "innumerable robberies committed on the fields, irrigated lands, and even robberies of wood from fences." The governor set out a number of specific punishments, with the most severe, for theft exceeding a value of twenty pesos, being sent "to the ring of shame," receiving "the number of lashes the judge would consider of justice," and spending six months in shackles without "going home to sleep or to leave the jail." In 1772 Ripperdá issued another long, complex edict to govern theft of green corn, with penalties including the thief being "tied up to the gibbet for one hour in a public place with the ears of green corn hanging from his head." Troops from the presidio were also "wont to commit transgressions on the campaigns . . . stealing horses, mules, cattle, sheep and goats." Because the soldiers were commonly illiterate, their commanders were instructed to periodically read them "the rightful punishment" that they could expect when apprehended. Edicts in force in 1776 included a curfew at 9 P.M.; prohibition of "short weapons" such as a blunderbuss (a type of muzzle-loading gun) or large knife; prohibition of "festive entertainment with disorders" after curfew, especially "in suspicious houses"; banishment from the community of unemployed men who refused to find work; public whipping for persons of mixed race or a month in jail and a fine of twenty-five pesos for Spaniards convicted of robbery or similar offenses; confiscation of any clothes laundered in the villa acequia; confiscation of cows, hogs, or other unconfined animals endangering the acequia; and fines and jail for parents who allowed their children to roam the streets after the curfew.[11]

CONTRABAND

French-manufactured goods in Louisiana, lower in price and of higher quality than Spanish goods available south of the Río Grande, were a powerful attractant to the residents of San Antonio. The French in Louisiana, as well as the Spaniards living in eastern Texas, were eager to trade these goods to the Indians for skins and pelts. It was natural for them to extend their trading network to the Spanish missions and presidios in Texas, where there was a ready supply of cattle, dried meat, tallow, and hides. Spanish authorities in Texas recognized the benefits of trade and repeatedly petitioned their superiors in Mexico City and in Spain to license or otherwise legalize trade. In fact, the Spanish settlements in eastern Texas, so far from San Antonio and the interior of Mexico, had little hope of developing if they did not trade with the French. In the early 1750s the Spanish authorities, recognizing that trade with the French was already substantial, issued edicts to suppress it. Although in 1757 the viceroy permitted some limited sale of livestock to the French, the king of Spain himself prohibited it in 1760. Even after the French ceded Louisiana to the Spanish in 1763, trade was not permitted, because Louisiana was part of the Captaincy General of Cuba whereas Texas was in the Viceroyalty of New Spain. The Crown maintained a strong antitrade policy throughout most of the century. However, weak local enforcement and a continuing demand for European goods made smuggling a major part of the eastern Texas economy.[12]

Governor Ripperdá, though sympathetic to the logic of trade with Louisiana, was continually reminded by officials in Mexico City of his duty to suppress smuggling. In 1774 the governor of Spanish Louisiana suggested limited trade of Texas livestock for manufactured goods from Louisiana, but when Ripperdá requested permission from the viceroy, he was rebuffed. In February 1775 Viceroy Bucareli warned Ripperdá that "notwithstanding my orders to cut all correspondence with the residents and nationals of the government of Louisiana and to deny them entrance to [the presidio in San Antonio], as his Majesty has resolved, you still continue to trade with them and frequently permit their introduction [to the presidio]." In a subsequent letter, Bucareli complained to Ripperdá of "the illicit trade and lack of obedience of the citizens of the town named Nuestra Señora del Pilar de Bucareli" on the Trinity River and named, ironically, for the viceroy himself. The viceroy complained that the inhabitants of Bucareli lived "by their smuggling practices, for which they have much opportunity."[13]

Some San Antonio merchants appear to have obtained their contraband directly from Louisiana rather than to have depended on intermediaries. In 1774 Marcos Vidal, a San Antonio merchant, accompanied by Joaquín Benites and Nepomuzeno Travieso, took a mule train to Louisiana and bought a variety of consumer goods that he planned to smuggle into San Antonio. However, he was apprehended by officials in Bucareli, the contraband was seized, and he was sent to San Antonio for trial. But before Vidal could be tried, he escaped and fled to the south, drowning in an attempt to cross the Río Nueces. As was commonly the case, the contraband was inventoried and then sold at auction. Ten kinds of fabric, a total of over 110 varas in length, along with forty handkerchiefs and a few other items were sold for a total of 225 pesos, probably to one of the tailors in the villa. The fabric included crimson satin, German linen, striped muslin, chintz from Holland, colored spun linen, and fine flowered silk. Other contraband included one firelock, three secondhand flintlock muskets, two secondhand blunderbusses, four axes, six salt shakers, twelve table knives, two pots, ten metal door plates, one arroba of coffee, and two hundred packs of playing cards. The horse and seven mules Vidal used to carry his contraband were also sold. The total value of goods sold at auction, including the fabric, was 481 pesos.

In related litigation, Benites and Travieso were convicted of helping Vidal smuggle mules to Natchitoches, despite their pleas that they did not know that the mules in the train were contraband.[14]

The state held a monopoly, or *estanco,* on sale and taxation of tobacco and salt. The tax on tobacco was two and one-fourth reales per libra, and avoiding the tax was a serious matter. In 1775 Jacinto de Mora, while visiting the town of Bucareli, received two bundles of tobacco weighing about eight libras (one libra equals about 1.01 pounds) in exchange for a debt of five pesos. Mora returned to San Antonio, but he hid the two bundles "near a chapel on the other side of the river" before entering the presidio, where the governor himself searched Mora and his companions for suspected contraband. Later, Mora retrieved the tobacco and sold it for five pesos to Marcos Hernández. An informant soon alerted Ripperdá to the contraband, and Mora and Hernández were arrested. After obtaining confessions from the malefactors, Ripperdá ordered their houses searched and their possessions inventoried. Finding no more contraband, the authorities confiscated the tobacco, and Hernández was fined twice the value of the tax owed plus the jailer's fees. Mora, who was insolvent, escaped a fine but was sentenced to fifteen days in jail.[15]

In 1776 the Spanish governor of Louisiana sent a contingent of soldiers to San Antonio with goods that they hoped to trade for mules. Ripperdá refused to accede to the request because the king had "ordered that all relations with the people of [Louisiana] be cut," even though they were also "vassals of His Majesty." But as the herds around San Antonio multiplied, the pressure to trade livestock and hides for manufactured goods became stronger, and San Antonio's ranchers almost certainly found ways to circumvent the law, moving herds of livestock and shipments of hides and tallow to markets in both Louisiana and Coahuila.[16]

AGUARDIENTE, CHINGUIRITO, AND VINO

The inhabitants of San Antonio consumed substantial amounts of alcohol, and regulation of its production, quality, and sale was a particular interest of Governor Ripperdá (1770–78). The government set the prices of alcoholic beverages and prohibited their adulteration or dilution. *Mescal* and *pulque,* distilled from *maguey* (a kind of agave), were not popular in Texas as they were farther to the south in Mexico. Instead, the residents of San Antonio preferred aguardiente (distilled liquor made from grapes or sugarcane), *chinguirito* (a poor-quality liquor produced locally from sugarcane or grapes), and *vino,* or wine.

The principal legal source of wine and grape-based liquors was the region surrounding Paras, in southern Coahuila. Zebulon Pike described this area as the "vineyard of Cogquilla [Coahuila], the whole population pursuing no other occupation than the cultivation of the grape." Nearby, Hacienda San Lorenzo had fifteen stills, large cellars, and gardens "delightfully interspersed with figs, vines, apricots, and a variety of [other] fruits." In 1770, soon after he took office, Ripperdá proclaimed that he had been informed that "the noxious beverage called chinguirito is being sold at a lower price than the one established by the government." He ordered that stills "that might serve to make chinguirito or any similar beverages" be turned in to him within eight days or the owners would "immediately be fined one hundred pesos." But by 1774 Ripperdá's objectives had changed. Instead of prohibiting its manufacture, he was protecting the price and quality of spirits consumed by San Antonio residents. But he advised the populace that "introducing and selling spirituous liquors, especially aguardiente—either legitimately from Parras [Paras] or adulterated—at excessively high prices in this Presidio has been noted." Fines and other punishments were published for selling aguardiente from regions other than Parras, for selling aguardiente for more than six reales per half *quartillo* (a quartillo equals about one pint) or wine for more than four reales per half quartillo, for selling less than the approved measures, and for adulterating the product. Apparently the edict did not have the desired effect, and in 1775 Ripperdá had several houses searched for illegal liquor. These searches turned up several suspicious products, and citizens with apparent expertise in the subject were called on to examine the quality of the liquor "by mouth, by smell, by hand, and by fire." The aguardiente sold by Vicente Travieso was found to be the best of the lot—"legitimately from

Parras but . . . adulterated with a sixth part of water, at least." For this, Travieso was ordered to reduce the price from six to five reales per pint, even though he protested that he had not diluted the product "but had purchased it in such condition at the warehouse in the town of Santa María de las Parras." Joseph de la Santa's aguardiente was "muchly watered" with "no activity" and its price was also ordered reduced. The aguardiente of Juan Joseph Flores was pronounced "pure chinguirito," and that sold by Joseph Antonio Curbelo was declared to have been diluted with water and adulterated with piloncillo. Both were ordered poured out on the public plaza. By 1775 Ripperdá had become an active supporter of chinguirito production, writing to the viceroy that the beverage "could not be harmful and it is of known utility to its producers, as long as grapes are grown and if your Excellency approves it."[17]

DECLINE OF THE MISSIONS

By the early 1770s the number of Indians in the missions along the Río San Antonio had begun to decline. The reasons were complex. Some families fled the missions to resume their old ways of life. Others died of European diseases, such as smallpox, measles, cholera, and influenza. Because of the Indians' lack of resistance to such diseases, they suffered more than the Spaniards and those of mixed race. In addition, as Indians became acculturated, they left the missions to marry soldiers and civilian residents of the villa, as well as to live and work in the villa and on ranches in the region.[18]

Valero, which claimed 275 neophytes in 1762, had only 126 in 1772, and only 53 of these were in residence at the mission, the others being considered fugitives. The inventory of 1772 makes it clear that the mission had far more land, improvements, and equipment than its resident Indians could effectively utilize. The mission's three farms were "each about a league long," and all were "fenced in with poles." The Alamo madre acequia provided "plenty of irrigation." When the inventory was made in November, one of the farms was "planted in late corn," which was ripe and "should yield more than 400 fanegas [1,032 bushels]." This represented far more than the mission Indians could consume, almost three pounds of corn per day for each of the mission's fifty-three residents.[19]

Like its population, Valero's herds decreased between 1762 and 1772. For example, the number of saddle horses declined from 115 to 25 and brood mares from 200 to 33. From 15 donkeys and 18 mules in 1762, only 2 mules were reported in 1772. The number of sheep and goats decreased from a total of 2,300 to about 250 sheep and 1,500 goats. In addition, in 1772 Valero reported 80 hogs, 3 two-year-old colts, 9 two-year-old fillies, and 10 younger colts and fillies. Miscellaneous agricultural items included seven carts (down from twelve in 1762), fifteen saddle trees, fourteen bridles, two worn but serviceable harnesses, two branding irons, twelve pairs of spurs, forty sickles, two scythes, fifty-two plowshares, eleven hoes, fifty fanegas of corn, and twelve fanegas of beans. Other items on the inventory included three muskets, five shotguns, and approximately fifteen pounds of powder and some bullets. For processing cotton and wool, the mission had thirteen combs for wool or cotton, some large scales for weighing cotton, two spinning wheels, a loom, about twenty-four arrobas of cotton, twelve pounds of cotton thread, two and a half pounds of mesh or cheesecloth, ten pairs of wool shears, twenty-five arrobas of wool, and about one arroba of woolen yarn.[20]

Mission Concepción had also "suffered great reduction in the number" of resident Indians, from "58 families, numbering 207 persons" in 1762, to 140 persons in 1777, and 71 in 1789. Like those of Valero, its facilities exceeded the needs of its reduced population. The 1772 inventory noted that its irrigation dam, located in La Villita, was "completely made of stone . . . with an intake area of water that is made of stone and lime." Its acequia, described as "completed, about one league long," irrigated three fields north of the mission. "One of the fields harvests from nine to ten fanegas [one *fanega de sembradura* equals about 8.8 acres]; the other from three to four; and the third a little more than two." The first two fields had pole fences. The other was "enclosed by branches because it got broken down last year." The large field was estimated to "yield about 600 fanegas of corn [1,548 bushels, or about 18 to 20 bushels per acre]." Concepción's Rancho del Paistle consisted of fifteen leagues east and southeast of San Antonio between Cíbolo Creek and the Río San Antonio. The ranch had houses made of stone, but "they were abandoned in 1767 because of attacks by hostile Indians who took all the horses." The year the ranch was abandoned it had twelve hundred head of cattle and four hundred branded calves.[21]

In 1778 Father Morfí extolled Mission San José as "the first mission in America, not in point of time but in point of beauty, plan, and strength." He reported that San José's Indians were "well instructed and civilized and know how to work very well at their mechanical trades and are proficient in some of the arts." Except for "those who are daily brought in from the woods by the zeal of the missionaries," they spoke Spanish, sang well, were well dressed, had abundant food, and aroused "the envy of the less fortunate settlers of San Fernando, the indolence of many of whom obliges them to beg their food from these Indians who enjoy so much plenty and whose mission is in opulence." Despite San José's apparent wealth in 1778, its population of 183 neophytes was little more than half the 350 who had lived there in 1768. Its armory was used to store guns, bows and arrows, and lances needed for defense. Workshops held a loom on which "rich blankets, cotton cloth, sackcloth, and other heavy cotton and woolen cloth worn by the Indians" were woven. The mission also had a carpenter, a blacksmith, tailor shops, and "everything needed for a well-regulated community." The granary was built "of stone and mortar, with three naves and a vaulted roof," and should hostile Indians lay siege, "the besieged, having as they have their granaries well filled with food and plenty of good water in their wells, could afford to laugh at their opponents." The food was produced on the mission's farm, described as "about a league square and . . . all fenced, the fence being in good condition." Irrigation water from the Río San Antonio was distributed by the acequia "to all parts of the field, where corn, beans, lentils, cotton, sugar cane, watermelons, and sweet potatoes" were raised. The mission also had "a patch for all kinds of vegetables and there are some fruit trees, from among which the peaches stand out, their fruit weighing at times as much as a pound." San José's large El Atascoso ranch was west and southwest of the villa on both sides of the Medina and Atascoso Rivers.[22]

From 1762 to 1778 Espada's Indian population declined from 207 to 153. Its Las Cabras ranch joined El Atascoso on the west along the road from San Antonio to Laredo, which ran almost due south from San Antonio, and extended eastward to the Río San Antonio. It joined Valero's lower pastures (La Mora) on the southeast. In 1778 there were 26 people, including nine men, four women, and thirteen children, living at Las Cabras.[23]

From about 1,300 neophytes in 1756–57, the total population of the San Antonio missions declined to 709 in 1777, to 376 in 1785, and to only 213 in 1792. In 1785 Fray López attributed the decline of the San Antonio missions to epidemics, lack of help from the military in capturing and returning Indians who had fled the missions, and poverty resulting from the loss of cattle following Croix's decrees of 1778. Former neophytes found employment as vaqueros on civilian ranches, which increased in number following Croix's decrees. In addition, the 600 acres of irrigated farmlands distributed in 1777 and 1778 when the *labor de arriba* was completed must have provided additional employment.[24]

By the late 1770s the declining population of mission Indians had caused a significant decline in crop production, leading Commandant General de Croix to "implore and charge" the missionaries to plant enough wheat, corn, beans, and barley to supply the community. He reminded the clergy that "service to the King" required that their crops be "abundant enough . . . so a good surplus or store [would remain] for the beginnings of the next coming year." Unfortunately, in February 1779 Governor Cabello had to report that the barley and wheat could not be planted because the arrieros bringing the seed (ten cargas of wheat and one of barley) from Camargo had been attacked by Apaches, who had killed two of the drovers and "ripped the sacks," scattering the seed on the ground. Nevertheless, a small amount of wheat and barley had been sown, and Cabello was planning to build a horse-powered mill to grind it. But on the night of March 30 a "tremendous and furious storm" almost totally destroyed the crop. Fortunately, the corn crop was successful, and more than six thousand fanegas were harvested and stored in the mission granaries.[25]

As the vigor of the missions declined, so did the conditions of the villa and presidio. In 1778 Morfí described the presidio and the villa's fifty-nine stone and mud houses and seventy-nine wooden houses as "all poorly built, without any preconceived plan, so that the whole resembles more a poor village than a villa, capital of so pleasing a province." The streets were described as "tortuous" and "filled with mud the minute it rains." Morfí attributed this poor state of affairs to the isleños' control of "practically the whole city government." He called them "indolent and given to vice" and judged that they did not "deserve the blessings of the land." In addition, the military had not been able to improve, or even maintain, the presidio. The soldiers' quarters, "originally built of stone and adobe," were "almost in ruins. . . . A few swivel guns, without shelter or defense" were mounted on the "poor stockade," apparently providing little protection to the inhabitants.[26]

During this period of declining mission populations, the villa was home to a large number of poorly paid laborers, many of whom must have chosen to leave the missions for a life of menial labor in the villa and on surrounding ranches. For example, the census of 1792 reports only twenty-five farmers but 556 laborers, probably including servants, day laborers, tenant farmers who did not own land, and the landless sons of farmers and ranchers. The standard wage for farm labor and other manual labor was minimal, two reales (one-fourth peso) per day, which was often paid in kind. In contrast, vaqueros were typically paid one peso per day, probably reflecting the much smaller number of men with the skills and courage the work required.[27]

Livestock numbers declined in the last years of the eighteenth century, probably for a variety of reasons. The missions were in decline and unable to manage their herds. Indians and townspeople continued to slaughter cattle for food, and exports took a toll. In addition, droughts in the early 1780s and the severe winter of 1786 may have further reduced the herds. Governor Juan Bautista Elguezabal reported that by 1803 the province of Texas had no more than one thousand sheep, wool was "very scarce," and "those who have any send it to Saltillo in order to manage to sell it." There was "a lack of meat," and only "the semi-annual slaughter of buffalos which takes place in the months of May and October" was able to "relieve the misery" of the majority of families, who otherwise "would no doubt starve."[28]

In contrast to the herds at the San Antonio missions, the cattle herds claimed by the La Bahía missions increased from the 1750s to the 1770s. Espíritu Santo's herds grew from three thousand branded cattle in 1758, to four thousand in 1759, to about sixteen thousand in 1768, and to fifteen thousand in 1774. The number of unbranded stock, orejanos, was even greater and was estimated at forty thousand. Rosario also had large numbers of branded cattle, claiming one thousand in 1754, four thousand in 1762, and more than ten thousand in 1780. Morfí reported that 695 people lived at the presidio, Mission Rosario, and Mission Espíritu Santo in 1778. However, Croix's decree of 1778 permitted private ranchers to legally capture and export both branded and unbranded stock. In addition, Morfí reported that the neophytes at Mission Espíritu Santo had a "fondness for the barbarous customs of their forefathers and it is necessary for the missionaries to be continuously on the watch to guard against their inclinations." By 1781 Mission Rosario had few neophytes left "because the majority apostatized and fled to some remote river or shore."[29]

Secularization of the missions began in the 1790s. Some of their lands and equipment were distributed to the few remaining residents, and others were sold. In 1795, after the secularization of its missions, San Antonio was home to sixty-nine ranchers, sixty farmers, thirty servants, ten merchants, nine tailors, six shoemakers, six cart drivers, four fishermen, four carpenters, and two blacksmiths. After secularization, the town took over management of the mission acequias, and the rules established for the acequia madre were extended to them. Though water rights were normally transferred with land ownership, by the early nineteenth century irrigation rights were being sold apart from land, and unused waters (*sobrantes*) could be rented if the owner did not plan to use them.[30]

ATTEMPTED REFORMS

As the nineteenth century began, Spanish governors of Texas made several attempts to solve the problems related to uncontrolled slaughter, sale, and exportation of cattle. Governor Antonio Cordero y Bustamante (1805–8) developed a system of internal passports to help keep track of travelers moving about the province. This, he hoped, would reduce thefts of livestock, especially by those of "low birth and few obligations." Because unregulated slaughter of cattle had decimated the herds around San Antonio, Cordero established the first official *carnicería,* a combination slaughterhouse and meat market. Two-week licenses were issued on a rotating basis by the cabildo to ranchers that supplied the carnicería with animals. A tax of three reales per head was levied, and the cost of meat and bone was set at one real per four pounds.[31]

In 1808 a new governor, thirty-two-year-old Manuel María de Salcedo, arrived in Texas directly from Spain. Son of the former governor of Spanish Louisiana and nephew of Commandant General Nemesio Salcedo, the new governor attempted several much-needed reforms of ranching practices. Governor Salcedo quickly recognized that Texas could not flourish without increased trade. Trade through Veracruz and Saltillo was unrealistic due to costs imposed by transportation and intermediaries. Therefore, Tejanos traded with Louisiana, where manufactured goods were available at reasonable cost and there were markets for Texas livestock. Salcedo recommended opening a port on the Texas coast to allow Tejanos to develop legal export and import trade. Other recommendations included locating immigrants on lands vacated by the secularized missions. Though the lands had been distributed to a few Indians and Spaniards, they were much less productive than they had been when the missions were active. Unfortunately, these recommendations, like those of his predecessors, were rejected.[32]

His ideas rejected, Governor Salcedo turned his attention to the depopulation of cattle herds in the area. In late 1809 and early 1810, Salcedo issued a number of proclamations designed "to end cattle thefts and fraudulent sales that constantly occur." He set up eight *síndicos,* each composed of several ranchers, to be responsible and manage the ranching industry around San Antonio. Their primary responsibility was to assure that ranchers killed or sold only those animals belonging to them. When ranchers killed cattle, they were to send "a piece of flesh" (presumably an ear with the earmark) to the síndico to prove that it was from their herd. Sales were to be conducted in the presence of a representative of the síndico. The síndicos were to frequently visit ranches to discover violations, and they were to keep a register of all the people and the number of livestock at the ranches. Salcedo also issued pronouncements regulating relations between ranchers and their employees, including the number and roles of those charged with tending herds, both of ganado mayor and ganado menor. For example, a *caporal* in charge of four mounted men and two on foot should guard each two thousand head of cattle. Three men should guard ganado menor and keep hogs out of fields. Roundups should be held in November each and every year, and strong fences tied with leather thongs should be built. Those who stole wood from these fences would be jailed and made to repair the damage. In August 1810 a more restrictive ordinance sought to strengthen the public slaughterhouse's monopoly on butchering livestock for sale of the meat.[33]

After more than a century of Spanish occupation and about half a century of private ranching, how many ranches were operating in Texas? Although we do not have exact information, substantial information was gathered in 1809–10. More than forty ranchers from the San Antonio area participated in the eight síndicos ordered by Governor Salcedo in 1809, though it is unclear how many individual ranches they owned. Historian Jack Jackson has located 30 ranches between the Neches and Sabine rivers and several more just east of the Sabine for the same period (see Map 4). It is noteworthy that a number of the ranchers in eastern Texas had Anglo names, whereas that was not the case around San Antonio or south of the Río Nueces. Despite the turmoil caused by Mexico's struggle for independence from about 1810 to 1821, the number of ranches increased in the years immediately between Mexican independence and the Texas Revolution in 1836. Historian Andrés Tijerina estimates that by 1833 the number of ranches in the San Antonio–Goliad area had increased to 80. In comparison, the number in the Nacogdoches area had risen to 50. Both areas combined, however, had far fewer than the region between the Río Grande and Río Nueces, which probably contained more than 350 ranches, although many of these had been abandoned due to Indian depredations following the Mexican Revolution.[34]

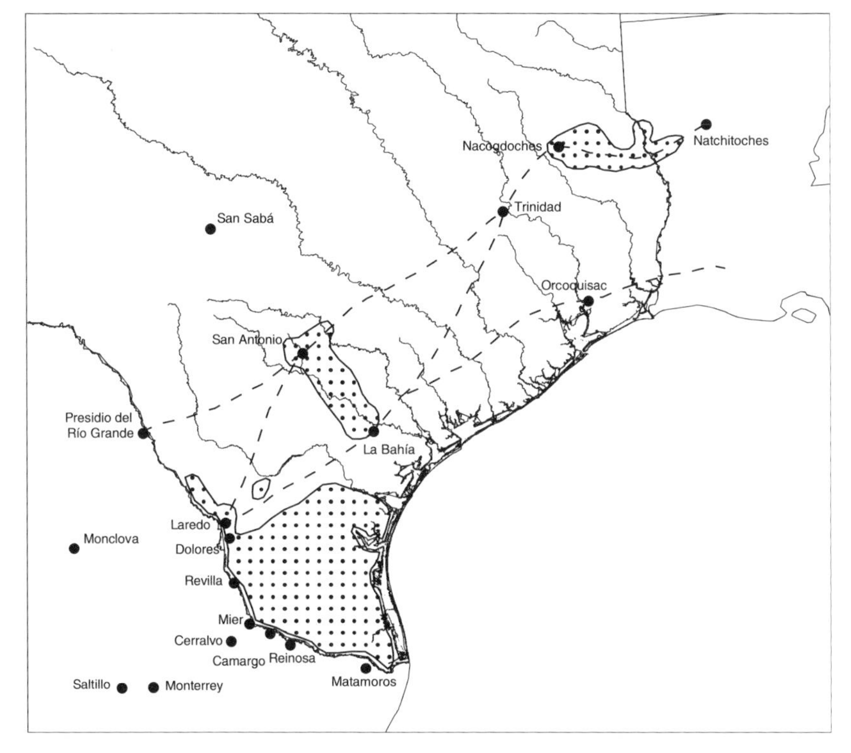

MAP 4. *Camino Real and ranching areas of South Texas, San Antonio, and Nacogdoches.*

STRUGGLES FOR INDEPENDENCE

Political conflict in Mexico, including Father Hidalgo's revolt in 1810–11 and Father Morelos's uprising in 1811–15, caused serious repercussions in Texas. Twice, in 1811 and 1813, Texas briefly fell into the hands of Mexican rebels. Reprisals by victorious royalists were brutal, draining Texas' resources. From 1818 to 1821 troops were withdrawn from the frontier, and ranches around San Antonio and between the Río Nueces and Río Grande became even more vulnerable to Indian attacks. In addition, bands of Anglo filibusters and other criminals took advantage of the political disarray to attack ranches in eastern Texas. As a result, many ranch families abandoned their properties and retreated south of the Río Grande or from eastern Texas to San Antonio or Louisiana. Census data suggest that a number of adult males died and others fled San Antonio as a result of the turmoil surrounding Mexican independence.

In 1820 San Antonio had more than two hundred widows, with more than eighty of those between sixteen and forty years of age. In contrast, there were only three widowers in that age group. In addition, there were more than eighty more married women than married men. Despite incursions by Anglo-American filibusters in 1816 and 1819, Texas remained under Spanish control until the summer of 1821, when it became part of an independent but economically depressed Mexico. In the 1820s and 1830s the newly constituted Mexican government faced numerous challenges. It was easy to neglect its northern frontier. As a result, from the Río Grande to the Río San Antonio the economy declined.[35]

After Mexican independence, Tejanos established new regulations to govern ranching and other activities. In effect, responsibility for local security was transferred from the national government to states and municipalities. Roundups were authorized from October to February. For animals rounded up and exported, taxes of one real per horse, two reales per mule, and four reales per head of cattle were charged. Livestock could not be freely bought and sold. To prevent unauthorized sales, a professional livestock trader, or *mercader,* was responsible for registering sales, along with the brands and numbers of livestock sold; collecting taxes; and delivering livestock to the town market or public slaughterhouse, the only location where cattle could be legally slaughtered. Another important official, the *jues de campo,* the rural counterpart of the town alcalde, supervised roundups, returned stray animals, settled disputes, deputized citizens, and pursued and prosecuted criminals. Local militias were responsible for defense, and the civilian *compañía volante,* or flying cavalry, was used to patrol the countryside, pursue Indians, and fight the Texians during the Texas Revolution. The Texas Rangers were later modeled after this highly mobile civilian cavalry.[36]

Between 1828 and 1834 Jean Louis Berlandier visited San Antonio several times. In 1828 it had a total population of 1,425. Because of Indian hostilities, which had only recently decreased, "there were to be counted in the entire municipality only 1,322 head of large livestock; 2,480 small livestock; and about 150 horses and mares." Berlandier described its inhabitants as "gay and not very hard working, and the dance is the chief amusement among the lower classes." Poor farmers occupied the mission lands, but they complained that hostile Indians made it too dangerous to cultivate the fields near the town. "At every moment the warring tribes kill some laborer, and they come almost constantly to steal the animals." As a result, agricultural production was inadequate, and the townspeople had to "go elsewhere, sometimes even to the Anglo-American colonies, to seek the grain necessary for their subsistence." Berlandier could

Cart

not understand why more crops were not planted on the "well-watered lands about the houses and the missions—even inside the presidio." He concluded that "moved by a principle of laziness," the farmers planted their corn in unirrigated fields "six or seven leagues from the dwellings (in localities truly exposed to attacks by the indigenes), solely in order not to have to take the trouble of watering the fields."[37]

Berlandier observed that the Texian colonies were "full of such animals as oxen, cows, horses, pigs, etc." and well supplied "with implements useful to their labors." In contrast, San Antonio farmers lacked even the most basic tools and equipment. Oxen were "the only animals used for draft and plowing" and were "badly harnessed" to carts with "wheels made of one or two joined pieces which have a lenticular shape." The rest of the cart was "nothing but an assemblage held together by ropes or rawhide; sometimes there is not even a wooden peg."[38]

Noah Smithwick, visiting San Antonio in the years just before the Texas Revolution, was also dismayed by the primitive equipment used in San Antonio. Lacking the tools and skilled artisans that the missions had provided in the eighteenth century, farmers in the 1830s were reduced to using plows that were "simply forked sticks, one prong of which served for share, another for handle, and the third for a tongue, which was tied on to a straight stick, the latter in turn lashed to the horns of a pair of oxen." The Mexican carts had "great, clumsy, solid wooden wheels."[39]

Berlandier noted that "corn yields with an astounding prodigality in Texas," and it could even "be exported if greater pains were devoted to its cultivation." Even though sugarcane grew "extraordinarily well in Texas," farmers could not hire sufficient labor to harvest it. The two piloncillo factories in San Antonio could not produce enough for local consumption, and sugar was imported from Louisiana, a cheaper source than Mexico. Even though Austin's colonists had been growing cotton for years, in 1834 only one or two farmers in San Antonio cultivated it. Wheat was not successfully cultivated in San Antonio. The varieties that had been tried had produced an abundance of leaves but little grain, and they were susceptible to rust. As a result, flour was imported from the United States by way of Louisiana, costing only one-third as much as Mexican flour. Berlandier predicted that wheat and grapes could be successfully cultivated in the hills west of San Antonio. He felt compelled, however, to point out that "more than a century after it was colonized the region remains static." San Antonio's population varied between about fifteen hundred and two thousand during 1805–33. However, its economy was declining dramatically in comparison with those of Nacogdoches, San Felipe, and La Bahía (whose name was changed to Goliad in 1829). All three were growing rapidly as a result of the influx of Texian settlers. By 1832, San Felipe was three and a half times the size of San Antonio. The mission acequias and irrigated fields gradually fell into disuse. After the Texas Revolution Indians took advantage of San Antonio's weakened defenses. In 1837 a visitor from the United States observed that the horse herds around San Antonio had to be watched during the day and driven into the town at night. Mexican herders seldom ventured "more than a mile from town on account of the Indians who often lurk in the vicinity, watching an opportunity to cut off their retreat and drive off the horses under their charge." A few days before his arrival a boy guarding a herd observed a Comanche creeping through the grass toward the horses. When the Indian "came within proper distance . . . he gave him the contents of an old musket." The Indian "with a loud yell . . . dodged into the mesquite thickets, where he was found a few days afterwards, nearly devoured by the wolves." In part as the result of Indian threats, by the mid-1840s Ferdinand Roemer reported that "although one can see small gardens next to the houses, one cannot find any corn or grain fields of any consequence outside the city limits." Though Mission Concepción's "artificial canals" were still there, its once fertile fields were "densely covered with mesquite bushes and trees," and the "fertile plain" that had once been "extensively cultivated" was "now a complete wilderness."[40]

Visiting Goliad in 1829, Berlandier was little impressed. It had, in addition to the families of soldiers at the presidio, about 650 inhabitants, "the greater part quite lazy, practically without industry, and whose labors most often can barely satisfy their needs." About thirty farmers, described as "little active and, above all, little hardworking" produced an average of five hundred fanegas of corn each year from their unirrigated fields. This amount of corn was inadequate for the needs of the town, so additional corn was purchased from Austin's colonists. But the government opened the port of Matagorda to trade, and the economy was spurred by trade moving inland in two-wheeled carts, each carrying eight barrels of merchandise. By 1832 the population of Goliad had rebounded to 1,439, but the impacts of this growth were apparently not all positive. "The greater part" of its residents had "a not very exemplary reputation," were "friends of every vice," including smuggling, and "all the ways of amassing booty [were] known to them." However, a cholera epidemic in 1833 killed 91 Goliad residents and caused substantial depopulation of the town, from about 1,600 before the epidemic to around 700 just afterward.[41]

Visiting the area again in 1834, Berlandier remarked that since Texian settlers had begun to arrive in the region, all the lands along the Río San Antonio between Goliad and San Antonio had been "measured and distributed." Though the law authorized "grants of only one square league, a number of individuals [had] acquired property of twelve to fifteen leagues." In addition, Berlandier was "pleased to observe" that since his last visit in 1828 "farming was beginning to be a promising enterprise" and "various newly built ranchos proved that the inhabitants of Texas are beginning to learn that agricultural products are the wealth which Nature has reserved for them." Jackson confirms Berlandier's observations with records that at least twelve land grants ranging from one to more than four leagues each were awarded between 1824 and 1834 in the Goliad area. At least three of those receiving the grants appear to have had Anglo surnames.[42]

From 1836 until 1845 both Mexico and the Republic of Texas claimed the land between the Río Grande and Río Nueces. After Texas joined the Union, both Mexico and the United States claimed the region. Rancheros living between the two rivers were put in a precarious position by the political uncertainty and violence the dispute created. The Mexican Army could no longer protect its northern frontier, and Texas had its hands full guarding the rapidly expanding frontier to the north. Comanches and Apaches were quick to take advantage of the situation. Rich in wild horses, ranch horse herds, semiwild cattle, and flocks of sheep and goats, the thinly populated ranching country between the Río Grande and Río Nueces was an inviting target for the marauders and for a decade after the Texas Revolution suffered frequent and damaging raids. Even though the more prosperous ranching families lived in stone houses designed for defense, workers in flimsy jacales, shepherds minding flocks, and travelers on lonely roads through the *brasada,* or brushlands, were easy targets. The Indians took advantage of the situation, raiding not only isolated ranches but also the towns along the river. The region around Laredo was particularly hard hit, with twenty-four deaths in the winter of 1835–36. Sheep and horse herds suffered devastating losses. In 1836 and 1837 large numbers of Apaches and Comanches ranged through the Lower Río Grande Valley burning ranches and killing livestock, as well as a number of people. On several occasions civilians and soldiers attempted to counterattack, with only limited success. After several years of small isolated attacks, in 1844 four hundred Comanches appeared in the valley, killing seventy people, taking numerous hostages, and burning several ranches. This time, however, pursuing soldiers overtook them and won several important battles, once killing ten Indians and rescuing over fifty hostages. This did not, of course, end the Tejanos' conflicts with the Indians, but these soon became far less important than raids by Texians hungry for land and cattle.[43]

By the mid-eighteenth century the missions along the Río San Antonio had attracted thousands of Indian neophytes, had developed a productive economy based on irrigated crops and cattle ranching, and had converted hundreds of Indians and their children into Spanish-speaking vecinos. A small civilian settlement had been established, led for many years by sixteen families of colonists from the Canary Islands. Irrigated farmlands worked by both the mission Indians and isleños provided the corn, beans, chilies, and other vegetables required by the community, as well as exported small quantities needed by missions and presidios without irrigable lands. Despite frequent harassment by the Apaches and Comanches, mission cattle herds had expanded beyond the mission Indians' ability to brand and manage them, creating chronic conflicts with both civilian farmers and ranchers. Trade, both legal and clandestine, had developed with settlements south of the Río Grande and in Louisiana, but mission herds had declined after Croix's decree of 1778. The neophyte populations of the missions also declined as mission Indians moved into civil society and the free Coahuiltecan bands disappeared. In addition, the fragile ranching economy between the Río San Antonio and Río Grande was rocked by the revolutionary tides of the early nineteenth century. This invited increased Indian depredations and settlement of Anglo-Americans (who called themselves Texians) on the fertile, unoccupied lands east of the Río Colorado. After first accepting Mexican citizenship, these English-speaking farmers and plantation owners rebelled, eventually defeating the armies of Mexico and establishing the Republic of Texas in 1836. Annexation of Texas by the United States in 1845 resulted in a war to settle the disputed boundary between Mexico and the United States. The Tejano ranching economy—buffeted by rebellion within Mexico, Indian depredations, the Texas Revolution, and the war between the United Sates and Mexico—would lack the power to repulse the influx of Texian ranchers and planters. The development of antebellum Texian agriculture and rural life, including its conflicts with Tejano ranchers, is the subject of part II.

Part II

THE TEXIANS

ANTEBELLUM FARMERS AND STOCK RAISERS

Chapter 6

GONE TO TEXAS

THE PERIOD FROM 1800 TO 1820 witnessed dramatic changes in both Spanish Texas and the southern United States. In 1803 the Louisiana Purchase doubled the size of the United States. The War of 1812 was followed by rapid westward expansion of the United States. Between 1800 and 1819 a number of American adventurers, such as Philip Nolan, James Wilkinson, Zebulon Pike, Augustus Magee, Samuel Kemper, and James Long, entered Spanish Texas, causing substantial disruption during a period of extreme uncertainty due to revolutionary activities in Mexico. Mexico, after great upheaval, gained its independence from Spain. With U.S. government support, farmers and planters forced the Cherokee, Chickasaw, Choctaw, Creek, and Seminoles to move off their lands in the southeastern United States westward toward Texas. Across the lower South the settlers developed small farms and plantations, grew corn and cotton until the lands were exhausted, then sold out and moved farther west onto new lands.[1]

AUSTIN'S COLONY

Mexican authorities, concerned about the rapid expansion of the United States, knew that they had to settle the almost-vacant lands of eastern Texas. In January 1821 the governor of Texas granted Moses Austin's application to settle three hundred families loyal to Mexico and the Catholic religion. But before he could accomplish his plan, Austin died. In the summer of 1821 young Stephen F. Austin, in possession of his father's grant, toured Texas from Nacogdoches to La Bahía (Goliad). In the aftermath of the Mexican Revolution, Texas was a depressed, depopulated, and vulnerable region. Many citizens with royalist sympathies had been killed or forced to flee. Austin was searching for a site to establish his colony. Nacogdoches, deep in the east Texas forests, offered few inducements. It had about three dozen inhabitants, a handful of houses, and a church. A few Anglo-Americans had entered the region and were squatting on Spanish land. Traveling the Camino Real, Austin found that Indians were killing travelers and stealing horses, even near San Antonio and La Bahía. He wrote in his journal that "the Spaniards lived poorly . . . have a few horses and cattle and raise some corn." However, below the Camino Real and between the Brazos and Colorado rivers, Austin found vacant land well suited for his colony. It received plenty of rainfall and had rolling tallgrass prairies dissected by numerous wooded streams; and its rich river bottoms could be cleared for the American style of plantation agriculture. He described it "as good in every respect as man could wish for, land first rate, plenty of timber, fine water—beautifully rolling."[2]

Austin located the capital of his colony, San Felipe de Austin, on the Brazos about eighty miles as the crow flies from the river's mouth. Chosen for its access to the river, its rich bottomland soils, and the surrounding prairies, San Felipe was soon the hub of a dynamic Texian farming and ranching community—on Mexican soil. Colonizing Texas with law-abiding families loyal to Mexico offered advantages to all, helping the Mexicans resist Indian raids and beginning to rebuild the economy. In addition, the large grants offered by the Mexican government were very attractive to mobile Texians. After 1820 the minimum price of public lands in the United States was $1.25 per acre. But in Texas, each farm family could receive one *labor* (177 acres), and each ranching family could claim a *sitio* (sitio de ganado mayor) or *legua* (4,428 acres; one square league). Naturally, most colonists declared themselves both farmers and ranchers and received both a league and a labor. Single ranchers could receive one-third of a league, and "families" composed of two or three bachelors could also be awarded grants. Colonists with substantial wealth who promised to import more colonists or slaves could receive additional land. For example, Jared Groce received ten sitios, and Austin, as *empresario,* was granted twenty-two.[3]

Austin and the Mexican government wanted only citizens of high moral character, and Austin specified that "no frontiersman who has no other occupation than that of hunter will be received—no drunkard, no gambler, no profane swearer, no idler." The majority of the Austin colonists were from the southern United States, moved to Texas for economic opportunity, and brought some sort of capital, including equipment, livestock, or slaves. Colonists were supposed to pay one *medio,* worth 12.5 cents or half a silver real, per acre. However, Austin offered generous credit and sometimes remitted payment, stating that he never turned a deserving person away simply because of inability to pay. Colonists were also exempt from customs duties for seven years and from general taxation for ten years. It is little wonder, therefore, that Austin was able to make 297 land grants to what have since been called the "Old Three Hundred" families.[4]

Most of the families Austin attracted to his colony lived on farms some distance from San Felipe, which Noah Smithwick described in 1827 as "twenty-five or perhaps thirty log cabins strung along the west bank of the Brazos River . . . the buildings all being of unhewn logs with clapboard roofs. . . . Austin's house was a double log cabin with a wide 'passage' through the center, a porch with dirt floor in the front with windows opening upon it, and [a] chimney at each end of the building." Other notable buildings were the blacksmith shop where Smithwick worked, Peyton's tavern, and the "saloon and billiard hall of Cooper and Sheaves [or Chieves], the only frame building in the place," Dinsmore's store, White's store, the Whiteside Hotel, and Godwin B. Cotton's newspaper office, which published the *Cotton Plant.*

Despite Austin's insistence on attracting only upstanding colonists, San Felipe had more than its share of scoundrels. They came for any number of reasons, but Smithwick thought that dueling and debt brought many. He remarked that "it was the regular thing to ask a stranger what he had done, and if he disclaimed having been guilty of any offense he was regarded with suspicion." Smithwick described cases of gambling, drunkenness, larceny, horse theft, assault, dueling, murder, and counterfeiting of coins, bills, and Mexican land grants in Austin's colony and surrounding areas. A popular verse is instructive.

The United States, as we understand,
Took sick and did vomit the dregs of the land.
Her murderers, bankrupts and rogues you may see,
All congregated in San Felipe.[5]

Despite its difficulties, the colony grew as more and more farmers, as well as a few merchants, doctors, lawyers, and artisans, took advantage of its rich, economical land and distance from past indiscretions.

J. C. Clopper reported that in 1828 near San Felipe there were "several planters already engaged in erecting sugar mills," and they planned to sell their sugar at ten cents a pound, cheaper than sugar from Louisiana could be sold. Many of the planters also had "cotton gins in operation, and the establishment of a cotton factory [was] already agitated." He reported that "here also is raised some of the fattest and most delicious beef and bacon in the world at no expense nor trouble, the grass of the prairies and mast of the bottoms makes it all." Salt needed for preservation of meat was "made abundantly and sold remarkably low." In addition, "the waters abound[ed] with the finest fish, oysters, crabs, and turtles; and buffalo, deer, and bear were plentiful."[6]

Austin boasted that in 1828 the colony contained about three thousand inhabitants, had "a number" of cotton gins and mills with "several more" under construction, and had produced about six hundred bales of cotton and eighty hogsheads of sugar. He claimed that "there probably is not at this time such an opening on the globe for industry and enterprise," and "men of large families and small or no capital cannot do better than to emigrate to this country." In Austin's colony he could "settle all his children around him" at a cost of no more than "four cents per acre, including surveying." Austin continued to promote the colony, boasting that in 1831 the colony produced "upwards of one thousand bales, of five hundred pounds of clean cotton each bale." He predicted that in 1832 production would "greatly exceed that amount." In 1833 Austin estimated that all of Texas produced six thousand bales of cotton. Cattle and hogs were so numerous that it was "difficult to form a calculation of their number." But because so many settlers were crowding into the state in need of establishing herds, livestock prices were good. Fat cattle weighing 500 to 750 pounds were worth eight to ten dollars a head. Hogs weighing from 200 to 300 pounds brought three to five dollars each.[7]

THE LAND THEY FOUND

Settlers attracted to Austin's colony found virgin soils, rich bottomland forests, extensive creekside canebrakes, and broad tallgrass prairies capable of pasturing large herds of livestock. The warm, moist climate and fertile soils were ideal for livestock and crop production. Wood was readily available for fences and houses, and wildlife was abundant. Elias Wightman observed that winter grasses furnished good grazing during the short time the perennial warm-season grasses on the prairies were dormant. The oaks, pecans, and wild peaches in the bottoms fattened the swine, which thrived "the year round, with no other expense than scattering a little corn . . . to keep them tame." Farther inland, there was "a much greater proportion of prairie" with forested river bottoms up to eight miles wide. Mesquite trees were numerous and were valued for their beans, "eight to ten inches long, very nutritious, on which horses fatten." Live oaks were "not infrequent, over little rills and in valleys." Between the Colorado and Guadalupe rivers the land was "quite hilly with post oak, black jack, and hickory timber, and many rivulets of fine water." Wightman predicted that in time "these prairies will be susceptible of fine sugar and cotton plantations" and "a boundless range for horses, cattle, sheep, and goats." Along the Spanish road connecting San Antonio and Nacogdoches were found timbered bottomlands, rolling prairies, and sandy hills supporting post oak, black jack, and wild grapes. Valuable pine forests grew near the Colorado and approaching Nacogdoches in the east. Between the San Antonio and Guadalupe rivers the prairies abounded with "turkies [*sic*], deer, wild cattle, buffalo, and mustangs; the latter in gangs, from fifty to three or four hundred." To the southwest, the plains between the San Antonio and Nueces rivers were described as having "rolling prairies, delightful mesquite vales[,] . . . good water and fine range for stock." Wightman reported that "the mesquite grass which abound here is superior to every other known . . . the game is plenty, deer and turkies [*sic*] at all times, and buffalo in the winter. Wild horses are seen in immense gangs in all directions."[8]

Most of the Anglo-American farmers and plantation owners who moved into Texas in the 1820s sought out the rich bottomlands along the Brazos and other rivers in the region. In 1836 John C. Duval observed that "nearly the whole of the bottoms on old Caney [Creek; in present-day Wharton County] was covered by an unbroken canebrake sixty or seventy miles long and from three to five in width." This "dense mass of cane, briers and vines, with here and there a scattering tree growing in their midst" grew on "exceedingly fertile" soil that would eventually "be converted into one continuous sugar and cotton plantation." Smithwick recalled that in the fall after frost had killed the tops, the cane was burned. In the early spring before the cane sprouts began to emerge, a sharpened stick was used to make holes in the soil, and a few grains of corn were dropped in. When the cane began to emerge among the corn seedlings, the suckers were simply broken over with a stick—all the weeding that the crop required.[9]

In 1828 Clopper described the Brazos bottom as "a low flat black rich soil from five to six miles wide [and] well timbered." In many places the bottom was covered by impassable canebrakes, and the "greater part" was sometimes flooded when the river overflowed, on average "once in three years" but "sometimes two or three years in succession." Of course, successful settlement of the bottomlands required that the fields be safe from floods, at least during most years. Felix Robertson, a surveyor, tried to reassure potential settlers, observing that the Brazos rarely overflowed its banks, and when it did, "not more than one third of the low grounds" were flooded. W. B. Dewees, after living in central Texas for a number of years, concluded that central Texas rivers overflowed about every ten years, but usually in winter or spring, causing little damage to crops. In 1823 the flood occurred in late January; in 1833 it came in May, and "crops were slightly damaged"; and in 1843 it overflowed in February, "and no injury was then done, with the exception of drowning of stock." Dewees concluded that, in fact, floods were beneficial because "as the rivers rise the water spreads over the land, and in place of forming a running stream and washing away the soil it stands still, and as it passes off leaves a sediment from three to six inches in depth upon the ground, which improves and enriches the land." In contrast with Dewees' optimistic assessment, those who had lived through the 1833 flood on the Brazos told of losing their corn in the roasting-ear stage and being forced to subsist on meat, curd, and sweet potatoes until the following spring.[10]

In the early 1850s Elise Waerenskjold wrote to her friends in the Old World that Norwegians who were not independently wealthy would be wise to move to Texas. "I believe Texas is the best of the states to migrate to, partly because the climate is milder and more pleasant than in the Northern states and partly because the land is cheaper." Waerenskjold noted that immigrants "could work themselves up in a short time to an independent position free of worries about the daily bread." Land, for those who wished to purchase it, could be obtained for thirty-five cents to two dollars per acre, but she feared that "this will not last long, as land is rising in price." However, "without owning land a person can here acquire as many cattle as he pleases" and pasture them on the abundant unfenced prairies.[11]

Most of the farm families who arrived in Texas in the 1820s and 1830s were from the southern United States, and only about one-third owned slaves. Many arrived with very little capital, having "worn out" one or more farms in Alabama, Mississippi, Arkansas, or another southern state. Without the labor or funds necessary to maintain their soil fertility with green manures or to collect and spread animal manures, they had been forced to leave and were seeking cheap, virgin soils on the frontier.

Although some progressive agriculturists saw abandonment of worn- out lands as a technical or moral failure, it usually made good economic sense. Throughout the South, land was cheap, especially on the frontier. However, labor was expensive, and all the activities needed to maintain soil fertility—herding and penning livestock, collecting and spreading manure, planting and incorporating green manure crops—required large amounts of labor. Therefore, small-scale farmers sold their worn-out farms and moved west to cheaper, more fertile lands. Large-scale farmers, or planters, invested in and even sought to multiply their supply of slave labor. Without enough land and family ties to keep them in place, the planters also moved west.

But farmers from the South were not alone in moving to antebellum Texas. In the 1840s and 1850s a number of Europeans left their depressed and sometimes repressive homelands to try their luck in Texas. Encouraged by early arrivals, many of them came to farm the rich soils of their adopted home. Of the European immigrants, the Germans were the most numerous and became well known for their farming skills. They came from a tradition that emphasized caring for and improving the land. As a result, their approach to farming differed from that of the small-scale southern farmer. This chapter describes how these two traditions met and functioned in antebellum Texas.

SMALL-SCALE SOUTHERN FARMERS

When a southern farm family moved to Texas, the first order of business was usually to plant a corn crop and hurriedly construct a crude lean-to or one-room cabin. In the mid-1820s Robert Hunter's family moved to the bottomland of the Brazos River in present-day Fort Bend County. After trading four cows and calves for about two hundred acres of land, they first built a primitive lean-to. As described by Robert, "Pa cut a big [ridge] pole 20 feet long, & put it up against 2 trees, & cut some long poles, & put one end on the ridge pole & the other end on the ground, & split out lathing & put on the poles & split out 3 foot boards, & covered it. It made a good house, & we lived in it 3 or 4 years."[12]

Later, when time could be spared, a more substantial log structure was built, sometimes with the help of hired craftsmen. The first step was to fell straight trees, usually of pine or oak, depending on availability. Logs fourteen to eighteen feet long were cut with an ax and then rolled or dragged with oxen to the site chosen for the cabin. The first course of logs was normally laid on rocks rather than on the soil to prevent rot.

One-room log house

If the builder had the time and resources, the logs were hewn to produce a square cross section. This allowed the logs to fit more closely together and reduced the amount of chinking required to keep out cold winds. Hewing also removed the bark, exposing heartwood and slowing rot. An ax was used to make a series of cuts down into the rounded face of the log to a predetermined depth. A foot adze, or sometimes an ax, was then used to split off the rounded portion down to the depth of the ax cuts, resulting in a relatively flat, uniform face the length of the log. The log was then turned, and another of the four faces was hewn in the same manner. After the logs had been squared, notches were cut at the ends where they joined the logs forming another wall. Several types of notches were used. A common notch, and the one almost always used with round logs, was the single saddle notch, a semicircle cut almost half the way through the log so that it would rest snugly on top of a log in the intersecting wall. Cutting the notch in the bottom of the log allowed the water to drain out of the cut and reduced the tendency of the corners to rot.[13]

More finely crafted log houses often had dovetail or box corners. This required that dovetails be cut precisely in the ends of the logs in order that the weight of the wall locked the logs together. When dovetail joints were used, the logs did not project beyond the corner, which was often faced with vertical boards to protect it from weather. In a survey of almost seven hundred log houses and outbuildings, Terry Jordan found eight distinctive types of notches used to lock the walls in place. In most cases, the logs were cut so that the weight of the wall rested on the corners rather than on the logs in the middle of the wall. This increased the stability of the house and reduced problems caused by drying and warping of the logs. The spaces between the logs were chinked with clay, adobe, or limestone plaster. If the cracks were wide, small sticks or thin laths split from logs could also be wedged in the cracks and covered with chinking material.[14]

Pole or hewn rafters were normally attached to the top log of opposite walls and joined together to form the ridge of the roof. The rafters, spaced about two feet apart, were joined to each other with long horizontal laths. Vertical studs were used to frame the gable ends of the roof. Roofing materials were usually either shakes or boards. A froe was used to rive, or split, shakes from cypress, white oak, or other straight-grained hardwood blocks. These shakes were usually about thirty inches long, fifteen inches wide, and one inch thick; and they were tied, nailed, or pegged to the laths. The roof often leaked at the peak, and the shakes on the side of the roof facing the prevailing wind were sometimes allowed to extend over those on the leeward side. As Taylor Allen recalled, boards were also used as roofing material. Cut from logs with an ax, froe, or whipsaw, they were secured with weight poles.[15]

Most houses had puncheon floors made of split logs or thick slabs smoothed with an adze. The puncheons could be set in sand or soil, or they could be pegged to logs laid below. Window and door openings, cut after the walls were in place, were framed with split or sawn boards. The doors were made of puncheons or boards and were hung on wooden hinges. They latched on the inside with a flat stick or bar pegged loosely at one end to the door and cut to drop behind a catch fastened to the cabin wall. A latchstring was tied to the free end of the bar and passed through a small hole in the door eight to ten inches above. When the latchstring was pulled from the outside, it raised the bar, allowing the door to swing inward. In frontier Texas, the latchstring was "always hung on the outside" to allow "anybody to enter and partake of the generosity and hospitality of the inmates." But when trouble threatened, the string could be pulled inside, effectively locking the door.[16]

One-room cabins typically had a single small hide- or wood-covered window located high on the wall near the fireplace. In this position it provided light in the area reserved for cooking and most other domestic activities. At night, light was provided by candles, pine knots, and the fireplace itself. The wide fireplaces were best constructed of rocks and lime mortar. However, they often had "hearths of timber and chimneys made of mud and held together with sticks." If the fireplaces were made of wood, heavy timbers were placed at the corners to support a lattice made of horizontal sticks. The lattice supported "mud cats" made by mixing clay with moss, horsehair, or grass and forming the material into rolls eight to ten inches long and one to three inches in diameter. These were woven between the horizontal laths and plastered with mud on the inside to protect the wooden structure from the heat of the fire. But such wood and mud chimneys were notorious for catching fire. As a result, they were not attached firmly to the wall of the house, allowing them to be pulled down and away if they caught fire. Smithwick recalled one occasion in which a fence rail had to be thrust between a house and its blazing stick-and-mud chimney to "throw the latter down."[17]

Chairs, stools, and benches were made of rough lumber and rawhide strips. Bedsteads were made in one corner of the cabin by placing the ends of two poles in large auger holes in the logs of the wall and the other ends in an upright pole that formed a bedpost. Narrow strips of rawhide provided support for skins of buffalo, bears, or deer tanned with the hair left on. The Allen family made most of the items needed for daily life on the frontier, including cloth, bedding, and harnesses. Sheep were sheared twice a year, and the family would "sit up late at night picking out the burrs and trash from the wool, then wash and card it and spin and weave it into cloth," from which clothing and bedding were made. The family made candles from tallow and beeswax. The hides from their livestock were tanned at "the Red Oak Ooze Tan Yards," and they made their own shoes, harness, and bridles. Cooking utensils and tableware were simple. Bowie knives, wooden forks and spoons, gourds, and tin cups were used at the table. For cooking they had "a coffee pot, frying pan, old-fashioned ovens, skillet and lids, and in the absence of these the old time hoecake and ashcakes were baked around the fire."[18]

INDIAN TROUBLES

Indians were a frequent threat to settlers in antebellum Texas, especially those near the farming and ranching frontier. Numerous firsthand accounts of specific encounters and battles have been written, and it is clear that both the Indians and the Texian settlers had abundant reasons to fear and mistrust each other. This book does not attempt to summarize or characterize the conflict, but readers may wish to consult J. W. Wilbarger's *Indian Depredations in Texas*. Published in 1889, it purports to be "reliable accounts of battles, wars, adventures, forays, murders, massacres, etc. etc." To keep the these memories alive, Wilbarger "carefully obtained from the lips of those who knew most of the facts" accounts of over 230 incidents, most of which resulted in one or more deaths of settlers, Indians, or both. Of these, only 7 occurred in the 1820s. However, in the 1830s the Comanche sought to take advantage of the civil disorder caused by Anglo-American filibusters and events leading up to the Texas Revolution. A total of 84 incidents were recorded in the 1830s. "The citizens had to do their own defending," and "all the houses built on Little [R]iver in 1835 were evacuated and the settlers from the falls of the Brazos had to retire, leaving plenty of empty houses." After the Texas Revolution, President Sam Houston (1836–38) established conciliatory Indian policies and reduced the number of Rangers. However, President Mirabeau B. Lamar (1838–41) took a much harsher approach. In 1839 the Cherokees were driven from east Texas in the Battle of the Neches. In 1840 the Comanches were defeated in three significant battles: the March Council House Fight in San Antonio, the Plum Creek Battle near Gonzales in August following the Comanche raid on Linnville and Victoria, and the destruction of the main Comanche village on the upper Colorado River in October. During Houston's second term (1841–44) he again attempted a conciliatory policy toward the Indians, establishing a series of trading posts on the western frontier. As Texas' population continued to increase in the 1840s and 1850s, the frontier pushed westward. Under pressure of the U.S. Army and settlers, the Indians retreated, and the number of incidents reported by Wilbarger decreased, to 60 in the 1840s and 24 in the 1850s. By 1860 the frontier of settlement extended from Maverick County on the Rio Grande north-northeast through Edwards, Menard, McCulloch, Coleman, Eastland, Young, and Archer counties, to Wichita County on the Red River. However, after Texas joined the Confederacy in 1860, the U.S. Army withdrew. Large numbers of

> men enlisted and left the state, leaving frontier settlements exposed. The Indians took advantage of the settlers' weakness, and the number of violent incidents recorded by Wilbarger increased to 47. In response, many of the settlers on the western frontier were forced to retreat eastward to the protection of the larger settlements.[19]

The first order of business for settlers after constructing a simple house was usually to clear and plant enough land to raise the corn and other foods needed to feed the family for the first year. Daniel Shipman's father moved his family onto a league of land on Oyster Creek, about twenty miles east of Houston, in 1825. Arriving with "a few head of cattle, hogs, and a very few horses," they built "quite a comfortable house" and then began to clear cane and wild peach using "grubbing hoes, axes and other tools." While the brush was drying, they "made and hauled rails and built fence." After burning the brush, the family planted corn and pumpkins, using a sharpened stick about four and a half feet long to "make a hole in the ground, waddle it about a little, pull it out and drop the seed in the cavity, and cover with the foot or stick." When the corn was "about ankle high," the family "would take hoes or sticks and knock the [sprouting] cane down, as it was tender and easily broken." After about three years the family was able to obtain "some good oxen" that enabled them to plow "the cane and other roots, and after that we were able to cultivate that Oyster creek land in style." With "as much good land and stock" as they needed, the family grew corn and cotton "and continued to live in our good old Texas style for many years." Though his memories may have been colored by the passing years, Shipman recalled that "those were the happiest days that we ever spent." They had "no taxes to pay, and as good lands as is in the world." They grew corn, potatoes, and pumpkins, "as many vegetables as we needed and as much cotton as we wanted." They "had plenty of fat cattle, horses, hogs, oxen," and though they lacked money, the family "did not need much of that." They "lived on the fat of the land . . . as gay as fighting chickens."[20]

Though many Anglo-Texans brought with them all the tools needed to start a new farm or plantation, others were poor and were forced to make what they needed. W. B. Allen was one of the first settlers of Fannin County, at the northern end of the blackland prairie near the Red River. Like most Anglo-Texan farmers of the period, he raised both corn and livestock. Far from a large town, the family had to make many of the tools needed on their subsistence farm. His first plow "was made on the bull tongue shovel order—at least a dozen pieces of scrap-iron, old horse shoes and wagon bed irons were used to make it, and the plow stock was chopped by hand, as well as the double and single- trees, from felled trees." The family's harness, ropes, and halters were made of "rawhide and hickory withes, and rawhide strips were used to hold the wagon beds and plows together."[21]

FENCING

Because there were large numbers of cattle, hogs, and wild game roaming at large, fences were needed to protect crops, gardens, and farmsteads. Recently arrived settlers made do with whatever fencing materials were easily available. In the mid-1840s Ferdinand Roemer reported that two families had erected temporary huts and planted their corn in a field enclosed in a "brush fence, made of felled trees and brush, as protection against the cattle roaming about." More substantial fences were built as time allowed. Zigzag or "worm" rail fences were the norm in eastern Texas. C. W. Tait specified that on his plantation rails should be ten feet long and split from trees at least one foot in diameter. "In making fences, lay the worm [the zigzag pattern of the fence] four feet and a half wide, make it five feet high, & then stake with Post-Oak or mulberry rails well set in, with a heavy rider." A "straight-rail" fence, also called "stake and rail" and "post and rail," consisted of pairs of posts (firmly set at intervals in the soil) with rails stacked between them. Anson Jones built "stake and rider" fences around the "house field" at Barrington plantation, and he had picket fences around his house and garden. On the rolling blackland prairie south of New Braunfels, a Tejano rancher named Flores had a "spacious yard . . . enclosed according to Mexican custom with a palisade of mesquite trees rammed into the ground, which were further connected with strips of raw oxhides."

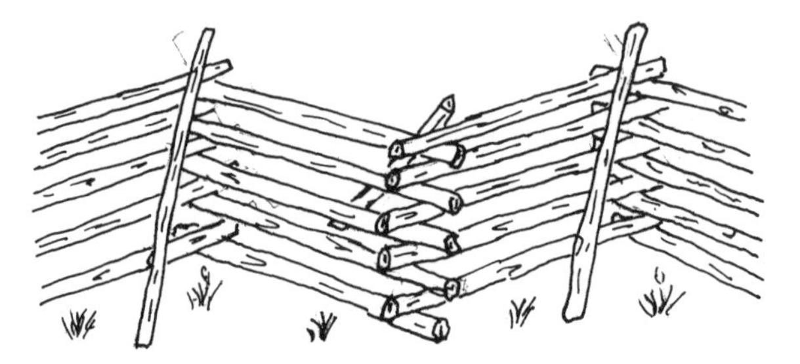

Zigzag rail fence

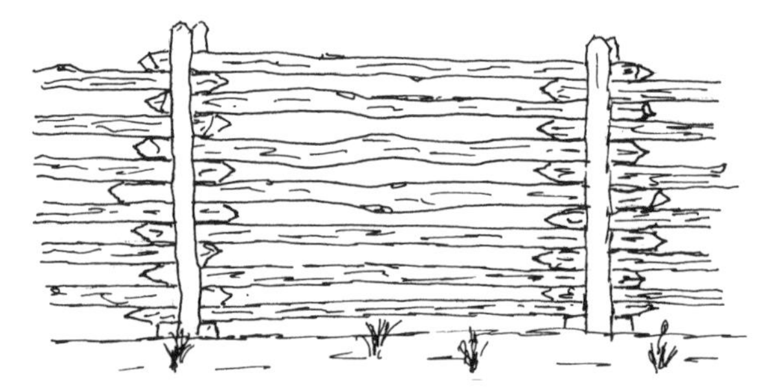

Straight-rail fence

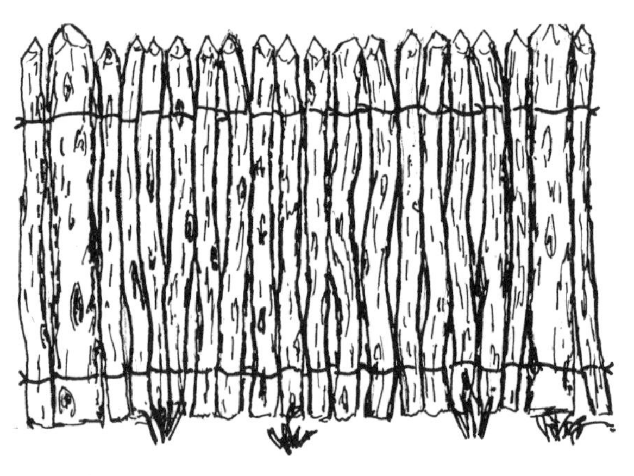

Palisade fence

By the 1850s wire fences were available, but Frederick Law Olmsted, a visitor from the eastern states, wrote that they were expensive "if well made." Near Goliad, Olmsted found an Irishman constructing "a novel sort of fencing of squares of turf piled into a handsome wall," indicating that "he had

> made many such here, and that they had proved strong and firm." German farmers in central Texas initially built zigzag, or worm fences, with cedar or oak rails. They also constructed Mexican-style picket fences and straight-rail fences with the rails tied or nailed between vertical posts placed side by side in the ground. By the late 1850s rock fences became popular with Germans in the Edwards Plateau. Miles of field and corral fences were built by stacking flat rocks without lime or cement to heights of three to five feet.[22]
>
> Thomas Affleck recommended planting hedges of Cherokee rose, which he reported "is very generally practiced in many parts of the country, and becoming more and more so each year." It was a "strong-growing, evergreen, running rose . . . with numerous strong thorns, growing readily from cuttings; and upon any land strong enough to yield twenty barrels of corn to the acre, will form an impervious fence in twelve years, if well cared for." In areas "where a planter is inclined to bestow a little extra care to form and maintain a neat, compact, and beautiful hedge," Affleck recommended evergreen thorn (pyracantha). However, it had to be "planted with care, using rooted plants, fully protected until solid enough to protect itself, and clipped at least once a year." Nevertheless, the diligent planter was rewarded with "a strong, solid wall of beautiful green, covered in spring with a blaze of snowy blossoms, and in the fall and winter with masses of bright scarlet berries." Bois d'arc, or Osage orange, was another plant used for hedge fences, and in 1848 Anson Jones hired someone to plant at least 850 feet of it on Barrington plantation.[23]

Many of the small-scale farmers who came to Texas were independent thinkers. They had been willing to move beyond the boundaries of the United States into an unstable foreign country. In fact, many left to escape the law or their debts. Some settlers often moved frequently, staying in one place only long enough to make a crop, then moving on. For example, in 1846 F. M. Cross's family moved to Cameron, "the county seat of the most western organized county [Milam]." At the time, the town had only a clapboard courthouse, a grocery store, and four families, one that served meals to travelers and another that had a steel grist mill where "every man had to do his own grinding." After buying two lots in Cameron and paying for them "with a brace of pistols that cost him $12.50 in Springfield Missouri," he built a log cabin in which the family lived during the winter and spring. However, Cross recalls that "a great many emigrants came into this part of the State and located on the Brazos and Little Rivers, so my father, desiring to go a little farther to the front, sold those two lots for $25 . . . [and] went on up the Little River and . . . built another log cabin and planted a little corn crop on Knob Creek, near the Pilot Knobs, and made a good corn crop without any fence" because they "kept the buffalo run off . . . [and] there was no other stock nearer than the three forks of Little River." Not content to stay on Knob Creek, after harvesting the corn crop, the family moved "nine miles above the three forks of Little River and settled on the Lampasas River" where they, along with Uncle Jack Nabors, "were the extreme frontier settlers on the Lampasas River." In all, Cross's father "bought and settled four places in the country before Bell County was organized." Of course, one of the reasons that families could move so readily to new land along the frontier was the availability of game. "As to wild game [in present-day Milam and Bell counties], there were buffalo, bear, deer, antelope and all kinds of smaller game from the lobo wolf down. The rivers were full of alligators and fish of all kinds."[24]

Some small-scale farmers had little ambition to improve economically and were content to subsist on very little income. Olmsted described a "squatter" living near Crockett in 1854. He owned no land; lived in a small log cabin; raised only corn and hogs; and sold a little hoecake, corn, bacon, and eggs to travelers. His only capital was a single male slave. His sales to travelers "furnished the necessary coffee and tobacco; nature and the Negro did all the rest." Another small-scale farmer living near San Felipe, who had come from Maine to Texas "because he was threatened with consumption," reported that his only cash crop was corn, which produced forty to sixty bushels per acre in a good year, but only ten bushels in a dry year. He employed four free laborers, two English and two German, and he was scornful of slave owners who lived without comforts in order to invest in more slaves. The rapid increase in the farm population and the need to purchase stock for new herds kept cattle prices high. Beef cattle were worth fifteen dollars per head, and working oxen sold for fifty to seventy-five dollars per pair. Sheep were almost unknown.[25]

Texans' attitudes toward farming and stock raising changed from one generation to the next. Olmsted reported visiting a rancher, a man of thirty who lived between Centerville and Caldwell, in 1854. The rancher recounted that his father had moved to Texas prior to the revolution with only a wagon, some horses, and his household effects. But he had prospered and was currently producing fifty bales of cotton a year, "equivalent to informing us that he owned twenty or thirty negroes, and his income was from two to three thousand dollars a year." However, the young man preferred ranching. "He was a regular Texan, he boasted, and was not going to slave himself looking after niggers." He had several hundred acres of prairie and woodland and "a large herd of cattle." He "could live as well as he wanted to, without working more than one month in the year." He told Olmsted that "for about a month in the year he had to work hard, driving his cattle into the pen, and roping and marking the calves." He and his neighbors helped each other, performing these chores "in a kind of frolic in the spring." "When he felt like it he got on to a horse and rode around, and looked after his cattle; but that wasn't work, he said—'twas only play." He also raised a little corn, which he ground in a steel grist mill. "When he wanted to buy anything, he could always sell some cattle and raise the money; it did not take much to supply them with all they wanted." The family clearly did not spend its money on its house, a one-room log cabin fourteen feet square with "battens of split boards tacked on between the broader openings of the logs." The roof was similarly ventilated, and "the sky could be seen between the shingles." The cabin had a "lean-to" made of boards on one side, but the door between it and the cabin, "sagging on its wooden hinges," could not be closed. The room was furnished with a "canopy-bed," a cradle, four chairs "seated with untanned deer-hide," and a table. "Crockery-ware," meal, coffee, sugar, and salt were stored in rough board boxes. And cooking utensils consisted of a "skillet or bake-kettle," a coffeepot, and a frying pan. The rancher also had a "rifle laid across two wooden pegs on the chimney, with a string of patches, powder-horn, pouch and hunting knife." Corn was kept in "a log crib at the horse pen," and meal was ground every day. Pork was stored in a log smokehouse.[26]

Clearly, not all farmers who did not own slaves were content with subsistence agriculture. Some were simply too poor to own slaves when they arrived but purchased them at the first opportunity. Robert Hunter's family owned no slaves when they moved to the Brazos bottoms in the late 1820s. Hunter recalled that after building a crude lean-to, they "went in to the cane brake & cut cane & cleared us up a field. We planted our corn with hand spikes & axes for 3 years before we could plow it, for the cane roots. We made from 40 to 60 bushels of corn per acre, & good big corn. After we got to plowing, we planted cotton. We made a bale & half of cotton to the acre. Pa sold his cotton 5 or 6 cents a lb." This cotton allowed

Hunter's father to begin purchasing slaves at a rate of about one per year. "Pa bought a negro woman, Ana, from old man Brown in Sanphilop [San Felipe], & then next year he bought another woman off Brown, Anas [Ana's] daughter Harriet. I believe it was 3 years, he bought 2 negroes Freeman & Seger, & the next year he bought a woman Mary, a African."[27]

FOODS

Most of the food Texians ate was locally produced. Corn and meat, prepared in numerous ways, provided the majority of the settlers' nutrition. Smithwick recalled that roasting ears were boiled, fried, and roasted—either "by standing the husked ears on end before the fire and turning them till browned all around" or by burying them "husk and all in hot ashes, the sweetest way green corn was ever cooked." When the corn had passed the roasting-ear stage, the still-soft kernels were grated off the cob, usually with a piece of tinware such as an old coffeepot, which had been "ripped open and spread flat on a board and punched full of ragged holes." The "large, soft, white Mexican variety" of corn grated easily and was made into bread that was "very rich and sweet, if a bit heavy." Dried corn was pounded into meal in a mortar made of a tree stump cut off three or four feet above the ground and "hollowed out by alternate burning and scraping till it would hold sometimes a peck of corn." The pestle was attached to the end of a long pole that rested in the fork of an adjacent tree, producing a lever that easily raised and dropped the pestle.

As the population grew in the 1840s and 1850s, hand-powered steel mills and water- or animal-powered gristmills were increasingly used to grind corn. The cornmeal was boiled to produce mush, to which milk, butter, molasses, honey, or sugar was often added. Mixed with cold water, cornmeal batter could be cooked in hot ashes to make "ash cake." The batter could also be placed on a board and propped near the coals to make "johnnycake." Spreading the batter on the blade of a hoe and cooking it over coals produced "hoecake," a welcome addition to the noon meal in the corn or cotton field. Coarsely ground parched corn was mixed with dried meat to make *piñole,* a staple of travelers who needed a light source of energy that would not spoil. Smithwick recalled that in the 1820s wheat flour was so scarce that "children forgot, many of them had never known, what wheaten bread was like."[28]

Because little wheat was produced in Texas until late in the antebellum period, wheat flour, shipped down the Mississippi River from the northwestern states, was often scarce and expensive. Coffee, imported as beans and usually drunk without milk or sweetener, was considered a necessity. Men both chewed and smoked tobacco; younger women dipped snuff while their older kin often smoked pipes. Large quantities of whiskey, brandy, cognac, gin, champagne, claret, and port were imported. Saloons were very common in most towns, and liquor was a common cause of violence and death.[29]

Meat was the other staple food of early settlers, who used beef, pork, and a wide variety of wild game and fish. Much of the meat was eaten fresh, but beef was often preserved by drying, and pork was commonly salted and smoked. The meat was cut into large pieces, placed in a trough or box, and covered with salt for about six weeks. The salt drew much of the moisture from the meat, which was then hung from the rafters of the smokehouse and smoked over a slow-burning fire to further dry and flavor it. The smokehouse was usually a tightly constructed log building ten to fourteen feet in length and width. Deer and bears were among the most prized game. In addition, both provided valuable skins, and bear grease was highly valued both as cooking oil and seasoning. As soon as possible, farm families began to develop gardens and orchards, both to diversify their diets and to bring a modicum of civilization to their frontier existence. Sweet potatoes were widely used, and pumpkins, cabbages, turnips, melons, and peas were not uncommon. Native pecans, wild grapes, and peaches were available in season. Sweeteners included locally harvested honey, as well as molasses and unrefined sugar produced on coastal plantations. In 1837 a traveler visiting a farm between the Brazos and the San Bernard was impressed with the "comfort and independence our host had prepared for himself and family." He had enclosed "some twenty or thirty acres of prairie upon the banks of the stream," where he was growing "potatoes, melons, most of the garden vegetables, and corn, all of which promised an abundant harvest. Sixty bushels of the last to the acre would be a fair estimate of the probable production."[30]

Waerenskjold, living in Van Zandt County halfway between Dallas and Tyler, wrote to friends in Norway that "very large and delicious" melons, pumpkins, and peas were planted intermingled with corn. She reported that "sweet potatoes yield very well if the summer is not exceptionally dry," but potatoes "of the Norwegian type" were "but little raised" and were eaten "as soon as they grow large enough in the spring." She felt that "it would pay to produce them . . . if they were left in the ground until fully grown." Vegetables did "very well" in northeastern Texas, and seeds were available for beans, peas, carrots, parsley, radishes, cress, lettuce, and fall turnips. However, vegetables requiring the long days and cool temperatures of northern latitudes could not be produced in Texas. Waerenskjold complained that she could not produce seed of "May turnips, any kind of cabbage or cauliflower, kohlrabe [*sic*], Swedish turnips, or French turnips which mature very early and would rot in August when the strong heat comes." She recommended that immigrants "bring along all sorts of seeds" and noted that "all these things can be raised without manure and with only the most perfunctory type of tillage." She missed the many fruit trees and berries of northern Europe, noting that "among fruits, the peach tree is the only one cultivated; it bears the third year after sprouting." However, wild plums were common, grapevines grew "in profusion everywhere," and persimmons were "delicious."[31]

Though most settlers had cattle, they were generally inferior milk producers except in the spring and early summer, when the prairie grasses were nutritious and plentiful. Waerenskjold wrote to her friends that in northeastern Texas "an overabundance of good hay grows wild" on the prairies, and "one only needs to cut and stack it to have butter and milk all winter." However, the majority of Texas farmers preferred "to do without milk and butter throughout the winter and let the cattle shift for themselves during this season."[32]

CURRENCY

Throughout much of the antebellum period, trade was severely limited by a scarcity of circulating coins and reliable paper currency. In the late 1920s, according to Smithwick, "Money was as scarce as bread. There was no controversy about 'sound' money then. Pelts of any kind passed current and constituted the principal medium of exchange." A few years later Smithwick became personally acquainted with counterfeit Mexican pesos made of silver-plated copper and counterfeit U.S. banknotes known as "Owl Creek money."

Immediately after the Texas Revolution in 1836, coins were available; however, during the worldwide financial panic beginning in 1837, worthless banknotes became common. In 1837 the Republic of Texas began to issue promissory notes as paper currency. These held their value for a time, but in 1838 they began to lose favor. From 1838 to 1841

President Mirabeau B. Lamar's government ran large deficits, and the value of its currency plummeted, reaching a value of only ten cents on the dollar. Governor Sam Houston, during his second term from 1841 until 1844, improved the government's financial stability. But in the absence of a stable Texas currency, throughout the late 1830s and early 1840s many types of coins and banknotes (issued by foreign governments, states, and cities) were exchanged at continually changing discount rates. Counterfeit bills were also common, and personal IOUs were used as exchange, even though the issuers sometimes refused to honor them. As a result, Texians were forced to barter slaves, livestock, cotton, skins, labor, and land for other goods. John Lockhart recalled that on one occasion J. W. McDade received a letter, but the postmaster refused to give it to him unless he paid the twenty-five cents postage due. McDade sought the money from his friends, but they had no cash. Finally, unable to locate twenty-five cents, he left his horse as security with the postmaster, took the letters, rode home behind a neighbor, and returned the next day with the coins necessary to redeem his horse.[33]

LAND TITLES

Proper land titles were extremely important to the development of Texas agriculture. Spanish and Mexican land grants had been awarded to numerous Tejanos and several empresarios before the Texas Revolution. Land speculators took advantage of the misconception that empresarios such as Stephen F. Austin had been given title to land rather than had simply received the right to place settlers on it. Speculators such as the New York–based Galveston Bay and Texas Land Company sold "land script," actually worthless and unnecessary permits to contract with the empresario for the land. Despite the confusion caused by sale of land script, Austin quickly awarded title to large areas of fertile lands. In July and August 1824 Anglo-American settlers patented fifty-six tracts in Brazoria County, the best bottomlands on the Brazos and San Bernard rivers and Oyster Creek. During the same period, almost all the best bottomlands in what would become Washington, Grimes, Austin, and Fort Bend counties were patented. In 1831 most of the land on the Navasota River and New Year's and Mill creeks were patented. Austin himself was awarded "premium" lands for his efforts. In 1828 he patented seven and two-thirds leagues (33,962 acres) of the most fertile "peach and cane" land in the Brazos delta. By the time of his death in 1836 he owned over 200,000 acres throughout the state.[34]

Several individuals received ten-league land grants in the Austin colony, and the Mexican government sold a number of eleven-league properties, often under suspicious circumstances. Smithwick recalled that in 1829 "an unscrupulous gang" of Texians obtained a number of genuine blank Mexican land-grant documents "lacking only the specifications [of the tracts of land] and the signature of the commissioner to complete them." After forging the signature of the commissioner, the gang established "a land office of their own . . . floating certificates to any amount of land for an insignificant consideration. Any good plug of a pony would buy an eleven-league grant." This sort of fraud and land speculation caused a great deal of resentment. As a result, the Texas Constitution of 1836 annulled many of the grants and took drastic steps to limit land speculation, especially by noncitizens.[35]

In an attempt to cope with growing confusion and speculation, in 1837 the Congress of the Republic of Texas ordered the newly formed General Land Office to examine and rule on the legality of land claims. The government recognized three classes of head rights: first class for settlers who arrived in Texas prior to October 1, 1837; second class for those who arrived between that date and January 1, 1842; and third class for settlers claiming land within the limits of colonies authorized by the republic but that did not meet the conditions of their contract. Heads of families and bachelors holding first-class head rights were entitled to 1,280 and 640 acres, respectively. Those holding second-class head rights were to receive half those amounts. Veterans of the army of the republic were awarded bounties ranging up to 1,280 acres, depending on the veterans' length of service. Additional land was given to disabled veterans and to the heirs of those who were killed in the revolution. Land scrips totaling over one million acres were also sold for fifty cents per acre in the early years of the republic. Applicants for land titles appeared before their county board of land commissioners and, if they could produce sufficient evidence of eligibility, were given certificates specifying the amount of land they were to receive. The applicant then arranged with a surveyor to locate and survey the land from that available in the public domain, usually giving the surveyor one-third of the land as compensation. The surveyors' notes were approved by the county or district surveyor and were certified by the General Land Office, which issued a patent for the land.

The process of obtaining title was slow and beset by numerous difficulties and irregularities. County land commissioners were sometimes dishonest and, even when scrupulous, were often besieged by hundreds of applicants, many with fraudulent claims. In addition, illness frequently slowed the progress of surveyors, and Indians threatened and killed them, especially when they tried to locate land beyond the limits of settlement. Nevertheless, the system worked. Firm titles could be obtained, and properties began to be subdivided, bought, and sold with regularity. For example, between 1846 and 1855 Julien Devereux expanded his landholdings in Rusk County from a few hundred acres to 8,253 acres. This land was composed of thirteen tracts ranging in size from 320 acres to 3,624 acres. Three of the tracts were from "leagues" or "grants"; the remaining were from "head rights." By 1855 Texas had received claims of 22.6 million acres of Spanish and Mexican land grants, 21.8 million acres of first class head rights, 8.6 million acres of second and third class head rights, 5.3 million acres issued as bounties and donations, 1.4 million acres issued for land scrips, and 1.0 million acres by special acts of the legislature. Over 6 million acres had also been promised to colonizers, land companies, universities, seminaries, and public schools. Of over 44.8 million acres of claims, the state had issued patents for 28.2 million acres and had twenty thousand claims for about 6.1 million acres outstanding.[36]

Throughout south Texas land titles were based on the original Spanish and Mexican land grants issued between the 1770s and 1830s; however, the Texas Revolution set in motion a series of conflicts that would, over the next forty years, destroy the very fabric of Tejano ranching. Many of the Tejano ranch families in the valleys of the San Antonio and Guadalupe rivers had been in Texas for a hundred years or more, and they had supported Texas during the revolution. Immediately after the battle of San Jacinto, Texians, many recently arrived from the United States, sought revenge for the massacres at Goliad and the Alamo. Blind to the fact that most Tejanos had nothing to do with the Mexican Army's behavior, reckless Texians threatened and attacked Tejano communities and families. In 1837 the old town of La Bahía was razed. In 1839 over one hundred Mexican families were forced from their homes around Nacogdoches. Most of these attacks were thinly veiled attempts to intimidate Tejano families, to force them to abandon their properties, and to seize their livestock and land.

In the 1840s many Tejanos attempted to secure Republic of Texas land grants. A total of 118 applications were granted around Nacogdoches, 1,228 along the San Antonio River, and 623 beyond the Nueces River. However, few were able to complete their patents on the land: 18 percent in eastern Texas, 25 percent along the San Antonio, and 68 percent south of the Nueces. The lower percentages near Nacogdoches and along the San Antonio were probably due to several factors, including ignorance of the law's requirements, greater Texian pressure to sell the land, and Indian raids. The ranches beyond the Nueces more successfully completed their patents because Texian raids were much less frequent in that area before the Mexican War.

Even though some Tejano families were able to perfect their titles, many were forced to give up the fight and leave. By the 1840s at least 200 Tejano families had left San Antonio. Many Tejano ranch families, dispirited by threats and rapid social change, sold out to Texians. From 1837 to 1842 thirteen prominent Texians purchased over 1.3 million acres around San Antonio from 358 Tejanos. But Texians were not alone in taking advantage of the turmoil. For example, 14 Tejanos purchased over 275,000 acres from 67 other Tejanos. By 1845 Texians had acquired forty of the forty-five Tejano ranches around Goliad.[37]

After the Mexican War, the violence against Tejanos increased. They were driven out of Austin in 1853 and 1855, Seguin in 1854, Matagorda and Colorado counties in 1856, and Uvalde in 1857. Also during 1857, masked gangs, probably made up of Texian teamsters, attacked and killed seventy-five Tejano teamsters. The "Cart War" effectively forced the Tejanos out of business and drove prices up by 30 percent before Texas Rangers began patrolling the San Antonio–Goliad road. The end of the Mexican War in 1848 brought rampant land speculation south of the Nueces River. Texians and Tejanos alike wanted legal land titles in a region that had been contested ground since 1836. In response the state of Texas established a commission headed by William Bourland and James Miller in February 1850. It was to examine Tejano land titles and make recommendations to the state legislature on how to perfect titles and facilitate land sales. They visited south Texas, collected a large number of original titles, and in August 1850 sent them by steamer to Austin. Unfortunately, all these titles, many over one hundred years old, were lost when the steamer sank. On the basis of duplicates, other evidence of title, and sworn affidavits, the legislature eventually perfected most of the titles. But their loss in the hands of the government increased distrust among Tejano landowners. The expense, difficulty, and uncertainty of the outcome led many Tejano ranchers to give up and sell their land to Texians at low prices. However, not all Tejanos sold, and in

1850–52 there were still 134 Tejano ranches in the brushlands north of the Rio Grande, with about 10 containing over 100,000 acres.[38]

GERMAN FARMERS

Beginning in the early 1830s a few German immigrants began to arrive in Texas. In 1831 Friedrich Ernst obtained a Mexican land grant in what would become Austin County. Over the next thirty years more and more Germans moved to Texas, often to join relatives who wrote home of cheap land and great opportunity for anyone willing to work. By the 1840s the Brazos, Colorado, and Guadalupe river bottoms had been claimed, and land was cheaper on the rolling prairies and woodlands

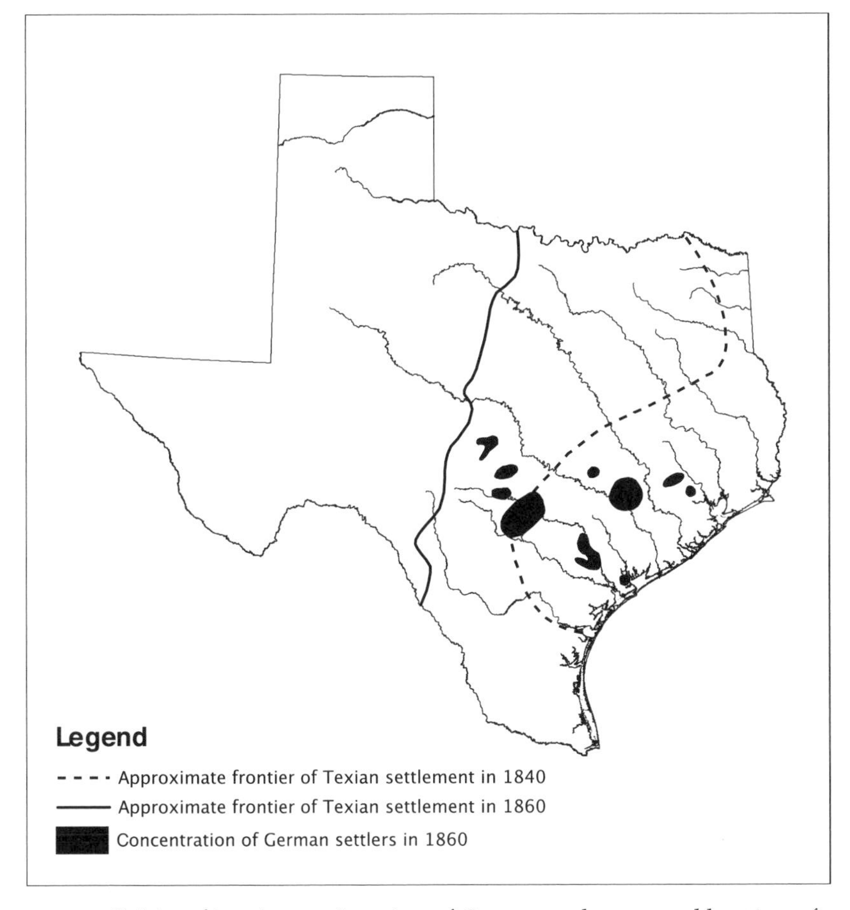

MAP 5. *Origins of immigrants. Locations of German settlements and locations of frontier in 1850 and 1860.*

between the rivers. As a result, in the 1840s and 1850s many Germans settled in and around the towns of Cat Spring, Industry, Brenham, Bellville, La Grange, Victoria, and Cuero. A second focus of German settlement developed to the west, along both sides of the Balcones Escarpment near the towns of D'Hanis, Quihi, Castroville, San Antonio, New Braunfels, Seguin, Mason, Fredericksburg, Comfort, and Sisterdale. Many Germans settled near these western communities between 1844 and 1846, brought to Texas by a German society known as the Verein, which collapsed in bankruptcy in 1847. However, German immigration continued, and by the late 1850s Texas probably contained thirty-five thousand to forty thousand residents who had been born in Germany or were children of German-born parents.[39]

Some Germans hired out as farm labor to Texian farmers and plantation owners. Others worked as carpenters, masons, and blacksmiths. Some pooled their resources to purchase land. By 1850 twenty-seven of thirty-eight German families living around Cat Spring owned farms. These numbers increased to seventy of eighty-five families in 1860. Initially, these farmers concentrated on subsistence crops, but they quickly discovered that cotton could be grown profitably, and 70 percent of the German farmers in Cat Spring were cultivating it by 1860. Most German farmers were averse to owning slaves, preferring to invest their capital in land and to hire free (or sometimes, slave) labor. Farm laborers were often other Germans or Czech speakers from Bohemia. Many of them worked for wages just long enough to purchase some land and become farmers in their own right. Over time, intermarriage and other ties brought the German and Czech communities closer, and most Texians simply regarded them as "the German element" or "the Dutch."[40]

Travelers often remarked that though they had initially been very poor, these German Texans were industrious and were increasingly prosperous. For example, the first German farmers Olmsted encountered between the San Marcos and Guadalupe rivers lived in small log cabins within fenced fields of about ten acres. He noted that their homes had "many little conveniences about them." The "greater variety of the crops . . . grown upon their allotments, and the more clean and complete tillage" of their crops "contrasted favorably with the patches of corn-stubble, overgrown with crab-grass . . . of the poor whites and slaves." Nearby, Olmsted found Germans living in "much more comfortable houses, boarded over . . . some of them having exterior plasterwork, or brick, laid up between the timbers, instead of boards nailed over them." These farms had larger fenced fields with "extensive crops" of corn and small areas, "not often of more than an acre," of cotton.

This "free-labor" cotton had been "judiciously cultivated," and in contrast to that on slave plantations, it had been picked "with care and exactness, so that none . . . had been left to waste."[41]

The farms observed by Olmsted along the Guadalupe between Seguin and Gonzales illustrate the difference between German small farmers and the planters who had come from the southern United States. The land on the east side of the river had been settled prior to the revolution by "extensive herdsmen and small planters." Most planters owned at least a thousand acres and between five and fifty slaves. Plantations typically had "a small front on the river, and extend[ed] back several miles over the upland prairie." The plantations were not fenced, and according to Olmsted, less than 1 percent of the bottomlands of the Guadalupe was cultivated. In contrast, the German farmers on the west side of the river had arrived after the revolution, and by 1854 they made up almost half the white population. They typically owned between twenty and one hundred acres, "which they till[ed] with their own hands." One planter who had moved from Tennessee to Alabama, Mississippi, and east Texas before establishing his plantation on the Guadalupe had high regard for his German neighbors, saying that though "there were some thieves among them . . . in general, they were very steady workers, trustworthy, and needing no watching when hired; they were very friendly-disposed people." Besides, they made "right fine" wine from the mustang grapes that grew in the river bottoms.[42]

CAT SPRING AGRICULTURAL SOCIETY

Wherever they settled, German immigrants formed close-knit farming communities with active social organization. In 1856 the German settlers of Austin County organized the Cat Spring Agricultural Society. With monthly meetings conducted in German, by 1858 its membership had swollen to over one hundred. The society sponsored lectures, subscribed to agricultural magazines, and conducted experiments with seeds ordered from the U.S. Patent Office. Members freely shared their experiences, recommending new crops and improved management practices. The minutes of their meetings provide valuable insights into the concerns of German-Texan farmers. Topics discussed in 1856, 1857, and

> 1858 included the marks and brands of member cattle raisers, protecting stored corn from weevils, fishing in area rivers and ponds, measuring corn for sale, improved peach varieties, use of corn soaked in lye to cure sows with worms of the spleen, whether cattle should be fed salt during wet weather, and how to prevent unlawful burning of the prairie. Lectures and discussions also dealt with bee keeping, care of peach trees, budding fruit trees, planting Irish potatoes, cultivation of hedges for fences, tobacco culture, keeping flies off livestock, cultivation of sweet potatoes, use of sugarcane suckers as planting materials, the cause of creamy-colored lint in cotton, methods of plowing "horizontal furrows" to prevent runoff and erosion, and producing sugar and molasses from sugar millet. Other topics included French chicken culture, drying peaches, well-drilling machinery, mowing machines, cane presses, corn and cotton planters, grape culture, wine making, irrigation, methods of breaking horses, building reservoirs to store water for cattle and for fish, planting hedges for fences, use of steam plows, sheep raising, and breeds of cattle. Possible cultivation of several unusual crops was discussed, including yellow and blue German lupines for green manure, sorghum, sweet sorghum, buckwheat, rye, turnips, chicken corn, lentils, and several varieties of peas. In 1857 the society placed an order to the U.S. Patent Office for red wheat from Turkey, white soft wheat from Tuscany, hard wheat from Algeria, Spanish peas from Argentina, lettuce from England, and lima beans, lentils, and barley from Peru.[43]

The climate of the Edwards Plateau around Fredericksburg was drier and cooler than that of the rolling prairies to the east, and it was known for its healthful properties. The spring-fed streams provided abundant freshwater, sites for water-powered mills, streamside timber, and narrow fertile bottoms for small-scale crop production. Olmsted described the region as "much broken, but well wooded and drained. In each little valley are one or more small prairies adapted to cultivation, and the hills are thickly covered with grass." Although the tall, coarse grasses afforded "poor nourishment in winter," cattle could find grazing "during the cold, in the river bottoms, where there is always some verdure as well as protection from the wind." The soil was described as excellent, with corn the principal crop, yielding "thirty to sixty bushels, from what would be considered at the North a very small outlay of labor." Wheat had been so successful "as to induce the settlers to send for harvesting and thrashing-machines." But hogs, "which increase with remarkable rapidity, and pick their living from the roots and nuts of the river bottoms," were one of the greatest sources of profit, and a "few ears of corn at night brings them all every day to the crib." Tobacco was cultivated by the settlers for their own use. Olmsted stayed in a German settler's house "upon a prominence, which commands the beautiful valley in both directions." He cultivated sixty acres below the house and "had produced 2,500 bushels of corn, besides some cotton, wheat, and tobacco" with only his labor and that of his two sons, aged fourteen and fifteen.[44]

By the 1850s the frontier of settlement had reached west of San Antonio to the lands along the Medina River. Castroville and D'Hanis had been established in the mid-1840s, their Alsatian and German settlers attracted by "a rolling sheet of the finest grass, sprinkled thick with bright, many-hued flowers, with here and there a liveoak, and an occasional patch of mesquit [mesquite] trees." The settlers had suffered a great deal during 1846, their first year, having arrived too late "to plant corn to advantage, and not having had time to make sufficient fences" to protect their crops from deer. The second year a hailstorm destroyed their crops, and "they lived on game and weeds, for the most part," even eating the rattlesnakes that "were then common about the settlement." A farm family at Quihi had arrived from Hanover in 1851 and was well established, having built a small cottage "composed of very simple and inexpensive materials, but . . . with more conveniences of living in comfort than many wealthy slaveholders' habitations." In 1853 they cultivated over fifty acres of corn and sold the grain to the U.S. Army for a dollar a bushel. They also had fifteen cows, four mares, and fifty hogs. By 1854 Castroville had about six hundred residents, with several hundred more living on farms nearby. D'Hanis, twenty-five miles west of Castroville, was a village of "about twenty cottages and hovels" with walls "of poles and logs placed together vertically, and made tight with clay mortar, the floors of beaten earth, the windows without glass." The roofs were "built so as to overhang the four sides, and deeply shade them." They were thatched with "fine brown grass, laid in a peculiar manner, the ridge line and apexes being ornamented with knots, tufts, crosses or weathercocks." With an average wealth of eight hundred dollars, mostly in cattle, the German settlers lived "in greater luxury than most of the slaveowning Texans," feeding Olmsted and his companions "venison, wheat-bread, eggs, milk, butter, cheese, and crisp salad."[45]

Terry Jordan analyzed U.S. Census data for both German and southern white Texians. For the eastern counties along the lower Brazos and Colorado rivers in 1850, less than 10 percent of the Germans owned slaves. In contrast, over 40 percent of southern white farmers were slave owners. Comparing only farmers without slaves, on average, Germans had smaller farms (175 vs. 321 acres), with less cultivated acreage (19 vs. 24 acres), fewer swine (18 vs. 39), horses (3 vs. 6), and oxen (4 vs. 6) than southern whites. In addition, on average, the Germans produced less corn (237 vs. 263 bushels), cotton (3.2 vs. 5.5 bales), and sweet potatoes (62 vs. 117 bushels) than their southern white counterparts. Jordan also discovered differences between the non-slave-holding German farmers in the eastern counties (Austin and Fayette) and their counterparts in the western counties (Gillespie, Hays, Guadalupe, and Comal). In 1850, on average, farmers in the western counties had smaller farms (166 vs. 175 acres) but produced more corn and cotton than those in the eastern counties (462 vs. 237 bushels and 3.5 vs. 3.2 bales). They also produced fewer sweet potatoes (40 vs. 62 bushels) and owned fewer swine (14 vs. 18), horses (2 vs. 3), milk cows (7 vs. 9), and beef cattle (11 vs. 15) than their eastern counterparts.[46]

In 1856, impressed by the prosperous German farmers that he found around San Antonio, Sisterdale, and Fredericksburg, Olmsted conducted an economic analysis of both livestock and cotton production in the region. He concluded that raising cattle and sheep was much more profitable than using slave labor to produce cotton. An immigrant with $3,500 to invest in a 1,000-acre ranch would have the following initial costs:

1,000 acres of well-watered land at $2.50 per acre	$2,500
House and furniture	750
Fencing and plowing 50 acres	500
Horses, oxen, and tools	350
200 stock cattle, $9 each	1,800
650 Illinois ewes, $4 each	2,600
Improved bucks	500
	$9,000

Olmsted projected the following annual income:

23 cows, $20 each	$460
23 steers, $20 each	460
600 lambs, $4 each	2,400
1,300 pounds wool, $0.25 cents per pound	325
	$3,645

Costs would include the following:

Wages for two farm laborers and two shepherds	−$720
Interest on $5,500 at 8 percent	−440
	−1,160
Net return on investment	$2,485

Thus, Olmsted calculated that with "average luck" the stock farmer could expect a net return of between 25 and 30 percent. In contrast, he estimated that investing the same $9,000 in a cotton plantation, with expenses including land and six slaves, would produce a small loss.

On a smaller scale, an immigrant with $1,000 to invest could purchase the following:

160 acres of land at $2.50 per acre	$400
Cabin and furniture	150
Tools, wagon, and cattle	150
Cows and pigs	150
Temporary subsistence	150
	$1,000

The first year the immigrant family would build the cabin and plow and fence a 20-acre field. Income would consist of about $250 from sale of 500 bushels of corn and $50 from sale of butter and pork. The family's living expenses would be about $200, and the remaining $100 would be invested in sheep. The second year an additional 10 acres would be plowed, and income from the corn crop would increase to $425. Income from butter and pork would increase to $75, and after setting aside $200 for living expenses, there would be $300 to invest in sheep. In the third year corn income would increase to $600, and the flock of sheep would continue to increase. Over a period of ten years, Olmsted calculated that the immigrant would have "a valuable farm, with herds and flocks which will insure him a comfortable subsistence without other personal labor than supervision."[47]

Though impressed by the efforts of German settlers, Olmsted had little positive to say about Texian agriculture, calling it "almost as rude and wasteful as it is possible to be." Crop rotation was not practiced so that "upon the same field the same crop is repeated, until all elements of yield are exhausted, when a new area is taken for the same process." Olmsted felt that "with cotton as the only export, and slaves as the only labor, no better system will ever be adopted." The profits of plantation agriculture were invested "in almost nothing else than slaves," and no consideration was given to improving the land "so long as virgin soils are at hand."[48]

SICKNESS AND HEALTH

Examination of diaries and other accounts of daily life in antebellum Texas reveals that diseases were common and greatly limited both the length and quality of life. Malaria, yellow fever, cholera, smallpox, measles, tuberculosis, dysentery, and other diseases took their toll on the population. The oppressive heat of summer was thought to cause the body to "lose its tone," causing "both mind and body [to] sink into a state of debility and indifference." Malarial fevers, or "intermittents," increased during the summer, and because the body had "lost much of its stamina [were] extremely difficult to eradicate." During the late summer and fall "the poisonous principle of the atmosphere" was felt to become "more highly concentrated," causing diseases "of a much more malignant character," including "remittents of the most dangerous types, cases of scarlet fever, neuralgia, every disease, indeed, dependent on miasmata."

In 1830 the government of Coahuila y Texas ordered an immunization program to stop the spread of a smallpox epidemic. In 1832 a traveler cured his recurring "fevers and agues" with "a small quantity of quinine" purchased for ten dollars from a young physician who had just arrived in Brazosport. Cholera swept up the Brazos in 1833, killing hundreds in Austin's colony before moving on to Goliad and Victoria, where it killed ninety-one and twenty-five people, respectively. Many in Goliad and other towns fled to San Antonio, hoping for a more healthful environment. Although the population of Goliad declined from over fourteen hundred before the epidemic to only about seven hundred in 1834, over the same period San Antonio grew from about sixteen hundred to twenty-five hundred. In the fall of 1837 a traveler reported that "great sickness prevailed in Houston, along Buffalo Bayou, as low down the San Jacinto as New Washington, and along the course of the Brazos." In addition, those living "upon the Trinity and in the eastern part of Texas had also their full share of disease and suffering." In 1853 Galveston was reported to have had more than two thousand cases of yellow fever with more than four hundred deaths after September 5. Physicians understood that a frost would slow the disease, though they did not realize that mosquitoes were the vector that cold weather suppressed.[49]

By the 1850s a variety of patent medicines were offered as certain cures. Browning's remedies were available through E. B. Wheelock & Co. Wholesale Druggists in New Orleans. For example, Browning's Cholera and Diarrhoea Remedy was advertised as a "never failing remedy for Cholera, Diarrhoea, Dysent[e]ry, Flux, and all afflictions of the Stomach and Bowels, accompanied with Cramps, Spasms, Purging, Vomiting, &c." Browning's Balsamic Expectorant was touted as "a safe and speedy cure of Coughs, Colds, Asthma, Bronchitis, Pains in the Side, Spitting of Blood, and all diseases of the pulmonary organs," including whooping cough. In cases of tuberculosis, it was advertised as producing "a radical cure, in all and every case, where the suffering will continue its use for a reasonable length of time; and even in the last stages of this dreadful disease, when the body is, as it were, dragging to the grave, it will give relief, and smooth its passage, though its course may not be stopped." Browning also offered Fever and Ague Mixture, touted as invariably curing "fever and ague, or intermittent fever, remittent fever, bilious fever, and dumb ague, all of which originate from nearly the same cause, namely, the miasmatic effluvia which arises from decaying vegetation in water or moist earth."

Others would not be outdone. "Dr. M'Lane's Celebrated Liver Pills" were touted as a cure for "Liver Complaint, or any Disease arising from excess of Bile on the Stomach," including cases of "Ague and Fever" and "Bilious Fever." "Dr. M'Lane's Celebrated Vermifuge" was advertised as a cure for sicknesses "produced by worms in the stomach and bowels." Francis T. Duffau's drugstore in Austin offered a number of extravagant claims for its medicines. Perhaps the most outrageous was for Holloway's Ointment, "a most miraculous cure" for "bad legs, bad breasts, burns, bunions, bite of mosquito and sandfly, coco bay, chiego foot, chilblains, chapped hands, corns, cancers, contracted and stiff joints, elephantiasis, fistulas, gout, glandular swellings, lumbago, piles, rheumatism, scalds, sore nipples, sore throats, sore heads, skin diseases, scurvy, tumors, ulcers, wounds and yaws."[50]

Planters used local medical doctors to treat a wide variety of illnesses and injuries, often running up substantial bills. They also purchased large amounts of medicine to treat both their families and slaves. For example, a typical list of items to be purchased by Julien Devereux from his agent in New Orleans included Moffat's pills & bitters, Brandette's pills, Bateman's drops, eye water, lip salve, calomel (mercurous chloride, used as a purgative), castor oil (a laxative), laudrum (probably laudanum, a preparation of opium), quinine (for malaria), morphine, rhubarb, aloes, camphor (to treat skin irritation), blister salve, epecac (ipecac, used to induce vomiting), peppermint, and paregarec (paregoric, an antidiarrhea drug). Home remedies were also widely used. Devereux loaned his wildcat skin to an acquaintance "to cure his wife's breast—recommended by Doctor Rains." Home remedies were also recorded in Devereux family records for hydrophobia, rattlesnake bite, cancer, fevers, lockjaw, coughs, and nervous headaches.[51]

Such was the lack of medical expertise that most deaths were officially attributed to symptoms rather than the diseases themselves. The following were the official causes of deaths in Austin in 1849: thrush, cold, teething, accidents, cronich, fever, dunat, croup, smallpox, congestive fever, numonia [pneumonia], inflammation of brain, inflammation of bowel, unknown, flux [diarrhea], hives, worps cough [whooping cough]. In 1860 Travis County had 100 deaths, of which 35 were infants less than one year old and 50 were slaves. The most frequent causes were reported to be pneumonia, croup, cholera infantum, teething, and pulmonary consumption. Other causes reported were poisoning, drowning, burning, murder, flux, bilious fever, bronchitis, indigestion, spasm, paralysis, epilepsy, yellow fever, typhoid, influenza, congestive fever, smothering, apoplexy, old age, rheumatism, pleurisy, cancer, whooping cough, cramp colic, inflammation, palsy, and dropsy.[52]

Small farms, either without slaves or with only a few, formed the backbone of the antebellum Texas economy. The farm families who settled them typically came from the southern United States, or sometimes from northern Europe, with relatively little capital. They usually struggled for a few years to establish their homesteads, but a great many took advantage of Texas' cheap, fertile lands to improve their financial positions. Most built better houses, many bought more land, and some began to acquire a few slaves. Others, drawn by the allure of the frontier or crowded by neighbors, moved repeatedly. Nevertheless, families with small farms made up the majority of the population. On the other hand, slave owners controlled a majority of the improved farmland and produced most of the cotton, corn, and sugarcane grown in the state.

Chapter 7

PLANTATIONS AND SLAVERY

WHEN STEPHEN F. AUSTIN ARRIVED IN TEXAS, he realized that the principal barrier to Anglo-American settlement was Mexico's prohibition of slavery. If it remained illegal, planters from the slave states would not move to Texas, and Austin's colony would languish. In an effort to secure the right for his colonists to introduce slaves, in 1822 Austin traveled to Mexico City, where the newly independent nation was being organized. The Mexican congress was in session, but before it could act on the slavery issue, it was dissolved by Agustín de Iturbide, who was proclaimed emperor. Under Austin's continuous urging, in January 1823 Iturbide's government promulgated a colonization law that permitted colonists to bring their slaves into Mexico. However, it prohibited their subsequent sale and stipulated that the children of slaves would become free at age fourteen. After Iturbide's overthrow in 1823, the congress prohibited the sale of slaves but did not prohibit settlers who already owned slaves from bringing them into the country.

In 1824 the state of Coahuila y Texas was proclaimed. Despite strong sentiment to completely prohibit slavery, the state constitution of 1827 did not emancipate slaves already in the state, decreeing only that the children of slaves would be free at birth, and after a period of six months no additional slaves could be imported. However, it was not long before slave owners moving into Austin's colony found an effective way of bringing their slaves with them—an 1828 law that allowed settlers to bring contracted "servants and day laborers or working men" into Coahuila y Texas. To take advantage of the law, before leaving the United States slave owners and their slaves signed a contract whereby the slaves agreed to accompany the owners to Texas, where the slaves would be freed. They would, however, compensate their masters by agreeing to pay their full value as slaves, the cost of travel to Texas, and the cost of their maintenance, all of which would be charged against their small annual wages. In addition, the slaves agreed that their present children, as well as any born in Texas, would serve their masters under similar terms. Neither the slaves nor their children would be able to pay off the debt to the masters; however, this was not considered slavery in the eyes of the government. It was indentured servitude or debt peonage—the same situation endured by poor peons throughout Mexico. The flow of slave-holding farmers into Texas increased with this assurance that they would not lose their valuable bondsmen.[1]

Despite the laws allowing indentured servants, some slave owners and traders sought to import bondsmen illegally. Before 1836 Leander McNeel, Ben Smith, James Fannin, and Monroe Edwards imported slaves from Cuba; Thomas League visited and probably imported slaves from Africa. When Texas gained its independence in 1836, President Sam Houston expressed his abhorrence of the continued importation of slaves from Cuba, and the Texas congress enacted a law prohibiting the importation of slaves from any country except the United States. However, in 1836 and 1837 Monroe Edwards, Sterling McNeel, and several others brought more slaves from Cuba. As Southerners flooded into Texas in the 1840s and 1850s, slavery became firmly entrenched, with fully one-third of Texas farmers owning at least one slave. Opposition to slavery, normally among immigrants from northern states or European nations, did not go unnoticed, and "the peculiar institution" had strong defenders. For example, the *Texas Almanac for 1858* declared that "every citizen of the United States should be the warm friend, the unceasing advocate and the bold defender of the institution of African Slavery." Among the reasons cited were that Africans were "inferior," were "incapable of self-government, or self-improvement," and could not "amalgamate with the white race without producing disease and death to the offspring." Second, "as a slave in a mild climate, the Negro is contented, cheerful, obedient and a long-lived laborer." Third, slavery was necessary to produce "the great staples" such as cotton and sugarcane. If slavery were eliminated, the "commerce of the civilized world" would be destroyed.

The result would be "anarchy, revolution, and internecine wars . . . [and] 'ruin' would be watch-word of every civilized State and nation."[2]

Slave-holding planters moved to Texas for several reasons. In many cases division of plantations among heirs left the sons of planters without sufficient land, often exhausted by years of cultivation, to sustain a plantation. Frederick Law Olmsted reported that by the early 1850s many of the plantations in western Louisiana had been worn out. Their owners had "gone to Texas," and "most of the remaining inhabitants live[d] chiefly, to appearances, by fleecing emigrants. . . . Every shanty sells spirits and takes in travelers. Every plantation has its sign, offering provender for sale." The inhabitants "obtained their livelihood by . . . shoeing the horses of emigrants . . . repairing the wheels of their wagons . . . selling them groceries" or selling corn and fodder for their horses.[3]

It was a long and difficult journey from the lower South, and by the time the immigrants reached eastern Texas, they were a weary lot. Olmsted described several slave-holding families traveling together in western Louisiana near the Texas border. "With fierce cries and blows" they drove their "jaded cattle." Covering ten or fifteen miles a day, they were led by a scout, "a brother or an intelligent slave, with the best gun, on the lookout for a deer or a turkey." Behind followed the "frequently ill-humored master," walking or on horseback, carrying a gun and urging on the black driver and his oxen. White mothers and babies, both white and black, rode in the canvas-covered wagons with the bedding and furniture. Following behind the wagon walked the "active and cheery prime Negroes, not yet exhausted," and bringing up the rear were "an old man, heavily loaded, with a rifle," an "old granny, holding on, by the hand, a weak boy—too old to ride and too young to keep up." And finally, "the stragglers . . . lean dogs or fainting Negroes, ragged and spiritless."[4]

Exhaustion of soils in the South and the promise of cheap fertile Texas soils resulted in rapid growth in the republic, and later the state, of Texas. In 1836 Henry Morfit, envoy of President Andrew Jackson, estimated that the new republic had 30,000 Anglos, 14,500 Indians, 5,000 Negroes, and 3,470 Mexicans. As migrants flowed into Texas, the population, both free and slave, increased rapidly. The state census of 1847 reported 102,961 whites, 38,753 slaves, and 295 free blacks. The U.S. Census of 1850 recorded 154,431 whites and free blacks and 58,161 slaves. Of this population, fewer than 13,000 lived in the five towns with populations greater than 1,000—Galveston, San Antonio, Houston, New Braunfels, and Marshall. Another 5,000 to 6,000 lived in towns with

100 to 1,000 inhabitants. Thus, in 1850 less than 10 percent of Texans lived in towns of at least 100 persons. About three-fourths of the free population, and even more of the slaves, engaged in farming, and about 86 percent of Texas farmers owned their farms. In 1860 the census reported 412,649 whites and free blacks, an increase of 267 percent since 1850, and 182,566 slaves, a 314 percent increase. Clearly, slavery was a thriving institution in the decade prior to the Civil War, with the percentage of slaves in the total population increasing from 27 percent in 1850 to almost 31 percent in 1860. However, because of the concentration of plantations on the lower Brazos, the percentage of slaves in the population was much greater there. For example, in 1850 and 1860 slaves made up just over 70 percent of the population in Brazoria County. In 1850 there were 3,507 slaves and 1,334 whites in the county; in 1860 there were 5,110 slaves and 2,033 whites living there.[5]

As the Texas population grew, so did its agriculture. From 1850 to 1860 the total number of farms almost tripled, from about 12,000 in 1850 to more than 35,000 in 1860. During the same period, the number of cattle increased by a similar amount, from a little more than 900,000 to 2.6 million head, and the number of hogs almost doubled, from 960,000 to 1.3 million. Production of corn almost tripled, 5.7 million bushels to almost 16.2 million bushels, but cotton production increased sevenfold, from about 58,000 bales to more than 400,000 bales. Wheat production skyrocketed, increasing more than thirtyfold, from about 42,000 bushels to more than 1.4 million bushels, as the central prairies from Austin north to the Red River were settled.[6]

The average age of Texas farmers was about forty years, and more than 80 percent were from the southern United States, with slightly more from the upper South than from the lower tier of states. Both in 1850 and 1860 less than 10 percent of farmers were from free states or were foreign born. As might be expected, farmers from the coastal prairies were more likely to have come from the lower South, whereas those in the blackland prairies and cross timbers were predominantly from the upper South. Foreign-born farmers tended to be concentrated in the western areas around San Antonio and in the rolling prairies between the Brazos and San Antonio rivers.[7]

In 1850 only about one-third of Texas farmers owned slaves. They were concentrated in three regions: the old Austin colony from the Gulf Coast inland along the Brazos and Colorado rivers, in east Texas around San Augustine, and in extreme northeast Texas. About 50 percent of the farmers in the southeastern coastal prairies owned slaves, compared to only about 20 percent of the farmers in the north-central counties between Belton and Dallas. Historians have generally used the term "planter" for a farmer with twenty or more slaves or who produced at least forty bales of cotton or forty hogsheads (barrels holding one thousand pounds) of sugar per year. In 1850 only 6 percent of all slaveholders qualified as planters, and 80 percent owned fewer than ten slaves. About two-thirds of the slaves in Texas lived on farms with ten to twenty slaves. By 1860 the number of slaves on large plantations had increased. Over sixty planters owned more than one hundred slaves, with four (all sugar planters) owning more than two hundred. About half the slaves lived on farms with ten to twenty slaves.[8]

In 1850 there were 466 planters who owned more than twenty slaves, and 375 produced at least 40 bales of cotton or 40 hogsheads of sugar. Ninety-one of these planters produced at least 100 bales or 100 hogsheads. The largest cotton planters in the lower Brazos counties were Thomas Coffee (500 bales), P. W. Cuny (432 bales), John Duncan (340 bales), Leonard Gross (300 bales), and Sterling McNeel (296 bales). The top sugar planters were David Mills (656 hogsheads), O. P. Hamilton (558 hogsheads), J. Greenville McNeel (511 hogsheads), Charles Powers (500 hogsheads), and Pleasant D. McNeel (420 hogsheads). By 1860 the number of planters producing at least 40 bales of cotton and 40 hogsheads of sugar had increased to 2,404; the number producing at least 100 of each had reached 872. A few prominent Brazoria County planters produced large amounts of both cotton and sugar, including Abner Jackson (622 bales, 586 hogsheads), David G. Mills (250 bales, 712 hogsheads), Aaron Coffee (600 bales, 120 hogsheads), and J. Greenville McNeel (275 bales, 300 hogsheads).[9]

As might be expected, slave-holding farmers typically owned more land and were wealthier than those who owned no slaves. Though they made up only one-third of Texas farmers, during the 1850s slave owners controlled about 80 percent of the property, 70 percent of the cash value of farms, and about 60 percent of the value of livestock. They also owned 70 percent of the improved farmland and about 65 percent of the unimproved acreage. With their lands, slaveholders produced about 90 percent of the cotton; about 65 percent of the corn, peas, beans, and potatoes; a little more than 50 percent of the wheat, rye, and oats; and almost 50 percent of the cheese and butter. In 1850 the "average" slave-holding farmer owned seventy-five acres of improved cropland, over one thousand acres of unimproved pastures and woodlands, and eight slaves. The cash value of the farm was about two thousand dollars, and total worth was almost nineteen thousand dollars. In contrast, the "average" farmer who owned no slaves had twenty-one acres of improved and less than four hundred acres of unimproved land. The cash value of the farm was less than six hundred dollars, and total worth was just over two thousand dollars. Planters in the southeastern counties were typically wealthier than those in other regions. By 1860 they owned an average of twenty slaves and had an average total wealth of more than forty thousand dollars. In comparison, slaveholders in the north-central wheat-growing areas owned an average of only six slaves and were worth, on average, less than fifteen thousand dollars.[10]

E. R. Wightman gave a good account of how a newly arrived planter could feed his family and slaves in the 1820s. Beginning in November "with a good team" and "one of the cast patent plows," the planter could plow half an acre of virgin prairie a day. This would allow him, in February, to plant thirty acres of corn. "Providing the sward was turned completely over" by the plow, weeds would not be a problem, and the corn would require "no more attention or working for the first year." The planter could expect his crop, planted early in the season on virgin land, to yield with "no risk" twenty-five bushels per acre, providing "corn enough for his use and some to spare." As soon as the corn was planted, the "all important sweet potatoes," yielding an average of five hundred bushels an acre, could be planted. For the slave-owning planter sweet potatoes could take the place of bread, and "when milk and butter can be had," slaves could "do good labor" on a bushel a week. In 1849 David G. Mills produced nine thousand bushels of sweet potatoes to feed his 344 slaves, and Abner Jackson produced nineteen hundred bushels for 285 slaves. In fact, Wightman insisted that he preferred to feed his slaves meals of sweet potatoes, gravy, and coffee, without cornbread or meat. Wightman also extolled the production of watermelons, pumpkins, muskmelons, cucumbers, figs, oranges, and peaches. Oats were reported to "produce beyond anything to be imagined." Pumpkins "spread over the ground to a surprising degree," and "where there are trees in the field, the vines run all over them, and you will see suspended in the air from the branches of the trees large pumpkins a foot through." However, rye and wheat would "be but little cultivated, as the more profitable crops of cotton and sugar will engross the attention of the planters and flour is offered to us on such easy terms from [New] Orleans, where the market is always glutted from the upper counties on the Mississippi and Ohio rivers."[11]

Planters establishing plantations on the virgin Texas prairies gradually broke more and more prairie sod and increased their croplands. Because the soils were sticky and the tough grasses had extensive root systems, breaking such soil was difficult. In 1854 near Austin, a prosperous

Texian plow

planter "with a plantation that would have done no discredit to Virginia" told Olmsted that the virgin prairie was usually "broken with huge ploughs, drawn by six yoke of oxen, turning a sod thirty-two inches wide and four inches deep." However, the planter felt that this practice buried the sod too deep, and he "thought it better policy to use a smaller plough, drawn by two or three yoke of oxen." This created a shallow furrow, and the sod decomposed more rapidly, facilitating subsequent tillage. In the *Texas Almanac for 1859,* Samuel Mather, a planter from Williamson County, agreed that virgin black and chocolate-colored prairie soils in his area should be "broken up in the first place about four inches deep, cutting the grass-roots about midway, which causes them to rot sooner than when the roots are turned up from the bottom." He recommended using a "prairie-plough," normally made of wrought iron and costing about twenty dollars. Four to six yoke of oxen worth thirty-five to fifty dollars per yoke were required to pull the plow. Sometimes wheeled plows were used. These did not require a farmer to guide them, only to "throw the plough out [of the furrow] with the lever for that purpose" at the end of the field, "then set it again for the next furrow." Such plows cut a furrow from twelve to twenty inches wide and could break from one and a quarter to two acres per day.[12]

Slaves were normally put to work early in the year plowing, harrowing, and planting. The crops were planted in sequence, beginning with those that were most tolerant of the typically cool, wet spring weather. In 1847 W. B. Dewees, who lived at Columbus on the Colorado River near the northern border of the coastal plains, recommended that tobacco seed and oats be planted on January 1, potatoes in February, corn between March 1 and 20, sugarcane in March, cotton between March 20 and April 15, and turnips in September. In addition, if sweet potatoes were put in beds in March, slips could be removed for planting throughout the summer. Dewees noted that river bottomlands produced two thousand to three thousand pounds of seed cotton per acre, whereas prairie lands yielded only fifteen hundred to two thousand. Nevertheless, he noted that "a man with the same force can raise as much produce on prairie land as the bottom land. The reason is this, the prairie land is so much more easily tilled, that one can take care of nearly as much again of it as they can of the bottom land." Average yields per acre of other crops were as follows: sugar, three thousand pounds; corn, thirty to eighty bushels; small grain, thirty to sixty bushels; tobacco, fifteen hundred pounds or more, with the second crop being nearly equal to the first. In addition, "when the season for cutting hay arrives, we go out into the prairies, select a place where the cattle have not been, and . . . cut as much as we wish."[13]

COTTON FACTORS AND COMMISSION FIRMS

At all the major Texas ports, merchants, often known as "factors," established businesses to purchase agricultural products from planters and farmers, as well as to sell them manufactured goods. Elliott Gregory, a New Orleans merchant, bought some of the first Texas cotton sold in that city, paying eleven cents per pound in 1831. But buying and selling Texas cotton was risky, especially when merchants also sold manufactured goods to the planters. Because money was scarce, transactions were often based on credit. Gregory found himself holding debt of planters Thomas Westall, who died of cholera, and Zeno Phillips, who went bankrupt. But Gregory's troubles created opportunities for Thomas McKinney and Samuel Williams, who organized McKinney, Williams and Company in 1834. The firm purchased a flatboat, the schooner *Brazoria,* and the steamboat *Yellowstone* to move cotton and merchandise to and from their business at the mouth of the Brazos. Though disrupted by the Texas Revolution, the firm remained in business, constructing a warehouse in Galveston in 1837, losing it to a storm, then rebuilding the warehouse, and adding a dock in 1838. In 1839 the firm brought the first vessel, the *Ambassador,* directly from England to Galveston with a load of merchandise. It carried the first load of Texas cotton directly to England without having to pass thorough New Orleans or an East Coast port. Despite the firm's successes in the 1830s, after Texas independence agriculture declined, farmers fell into debt, and merchants suffered. McKinney, Williams and Company also had the misfortune of losing at least three steamboats, and in 1842 it was taken over by another firm.[14]

From 1840 to 1845 Morgan Smith and John Adriance operated a general store in Columbia, fifteen miles from the mouth of the Brazos. They bought goods on credit in New York and sold them in their store as well as to other merchants farther inland. They also marketed cotton in New Orleans, New York, and England for some sixty planters. When Texas was a republic, planters were often in debt to their factors, and bankruptcy was not uncommon. As a result, one of a merchant's most important decisions was to whom credit was extended. Smith, who traveled regularly to New York to obtain merchandise, advised Adriance on which planters were creditworthy. In addition to their cotton warehouse, they obtained Waldeck plantation to keep the livestock, slaves, tools, and other goods collected in payment for planters' debts. Shipping was also risky. For example, Smith lost twelve to fifteen thousand dollars in uninsured cargo when the schooner *Eliza* wrecked at the mouth of the Brazos in 1842. The firm also suffered when floods damaged the cotton crop in 1843, and in 1845 the partnership was effectively dissolved. Smith took control of Waldeck and developed it into one of the most productive sugar plantations in Texas.[15]

Robert Mills was one of the most successful Texas merchants, beginning business with his brother Andrew in Brazoria in 1830. Selling simple manufactured goods such as cloth, clothes, shoes, and tools to area farmers, he purchased their cotton, Spanish moss, tallow, beeswax, hides, pecans, and other products. His goods even made their way south of the Rio Grande, paid for in Mexican silver coins and bullion. After Andrew died in 1835, Robert brought his younger brother David into the business. David operated the firm's plantations, and Robert bought from and sold to the farmers. After Texas joined the Union, Robert began to deal directly with cotton mills, traveling to Europe on steamships and arriving in time to sell his cotton to the mills before it arrived by sailing ship. In 1849 Robert moved the firm to Galveston and entered into a partnership with James and Edward McDowell. Operating in Galveston, New Orleans, and New York, the partnership prospered throughout the 1850s, and the Mills brothers were said to be worth millions in the late 1850s, before failing as a result of the Civil War.[16]

Julien Devereux's purchases for his Terrebonne plantation in Montgomery County illustrate the variety of goods that factors provided. In 1843 he purchased "1 Barrel of sugar, 1 Barrel of flour, 100 lbs coffee, 1/2 doz. Weeding hoes, 7 axes, some good spirits of wine, quinine, castile soap, candle moulds, looking glass, fine comb, and Black ink." In 1844 Devereux sold four bales of cotton weighing 2,319 pounds to J. Shackelford, Jr., of Houston for seven and a half cents per pound. Against the credit of $173.92 he purchased $135.87 of merchandise, including four sacks of salt, one hundred pounds of brown sugar, bagging, rope, a blanket, two axes, a barrel of Irish potatoes, cups, saucers, knives, forks, plates, half a bushel of onions, two pounds of pepper, garden seeds, rifle powder, eleven yards of calico, one pair of "Morocco" shoes, one pound of ginger, one pound of saltpeter, one pound of sulfur, twelve pounds of "Blester" steel, seven and a half pounds of tobacco, three gallons of vinegar, two gallons of whiskey, two pounds of bailing wire, and one barrel of apples. In 1845 he purchased $474.73 in goods from Paul Bremond in Houston, paying for them with the proceeds of eighteen bales of cotton weighing 9,263 pounds sold for five and one-eighth cents per pound.[17]

Advertisements in the *Texas Almanac* during the 1850s provide a glimpse of the variety of businesses serving the needs of Texas planters and farmers prior to the Civil War. For example, H. Runge and Co., Commission and Forwarding Merchants on Powder Horn Wharf in Indianola, sold "dry goods, clothing, hats, shoes, hardware, castings, iron, crockery, stoves, furniture, wooden ware, groceries, ox and mule wagons, buggies, coal, cement, lumber, building materials, &c." In Houston, Henry Sampson & Co. and John Dickenson advertised themselves as "Cotton Factor and Commission Merchants." W. R. Wilson dealt in "hardware, tinware, stoves of all sizes and shapes, pumps, scales, plows, mechanics' tools, agricultural implements, &c., &c." On the Strand in Galveston, E. S. Wood Hardware and Cutlery, under the "Sign of the Golden Plow," sold "agricultural implements, iron and tinware, iron and steel, wood and coal stoves, nails, castings, paints, oils, glue, and every article in the hardware line." Also on the Strand, Andrews and Gover provided "groceries, provisions, produce, liquors, glass . . . Galveston Bay oysters, in tin cans hermetically sealed . . . [and] a large and well selected stock of family groceries." Under the "Sign of the Cotton Bale" on Tremont Street in Galveston, Buckley & Byrne, advertising "quick sales—small profits," were "wholesale and retail dealers in domestic, plantation & dry goods." More discriminating planters might patronize P. & E. Reilley & Co. on Canal Street in New Orleans, who sold "French and British dry goods" and were "dealers in southern plantation goods." Wagons and carts to transport cotton, as well as wheelbarrows and "timber wheels" used to drag logs out of the woods, could be purchased at Phelps, Carr & Co. or David G. Wilson in New Orleans.[18]

MAINTAINING SOIL FERTILITY

By the mid-1850s many plantations in eastern Texas had exhausted their soils, and the owners had moved up the river or farther west. Olmsted reported that along the Sabine "the improvements the inhabitants have succeeded in making" were outweighed by "the deterioration in the productiveness of the soil." As a result, "the exhausted land reverts to wilderness." The region was filled with "many abandoned farms, and the country is but thinly settled." The inhabitants were "still herdsmen, cultivating a little cotton upon river-banks, but ordinarily only corn, with a patch of cane to furnish household sugar." Times were clearly hard for those settlers, who "didn't make corn enough to bread them" and whose "herds were in poor condition, and must in winter be reduced to the verge of starvation." Even their hogs were poor, "converted by hardship to figures so unnatural, that we at first took them for goats."[19]

Thomas Affleck was a writer, scientific agriculturist, and horticulturist. Born in Scotland, he studied science before moving to the United States in 1832. By 1840 he was junior editor of *The Western Farmer and Gardener* in Cincinnati, the beginning of a lifelong career practicing and writing about scientific farming. Familiar with the soil exhaustion throughout the South, Affleck was very concerned that planters were too shortsighted and failed to maintain their soil fertility. He observed that "a planter's almost sole wealth consists in negroes; his land is comparatively valueless." Nevertheless, Affleck strongly recommended that planters improve their soils. In 1859 he observed that although cotton prices were high, "but a very few years have elapsed since the markets of the world were glutted, and prices below the cost of production; and that the same state of things may exist again." He noted that "the few who, during these years of low prices, turned their attention to the manuring and otherwise improving their lands, instead of straining after large crops *because* prices were so low . . . [could], now that cotton commands such high prices . . . take an extra crop or two from their lands without injury to it." In addition, "land *in good heart,* it is well known, will sustain a crop during even disastrous seasons."[20]

Affleck understood that soil exhaustion was hastened by erosion, the removal of nutrients in the harvested seeds and fodder, and burning the crop residues to facilitate plowing. To counter these effects Affleck advised farmers to "beat down cotton-stalks and chop corn-stalks short, so that the plow may cover them up. Under no circumstances burn the latter; nor the former unless very large." He recommended thoroughly breaking the land in winter, turning under crop residues to promote their decomposition. He wrote that "no labor upon the plantation is better bestowed, than that which is applied to drawing the heavy cover of crab or crop-grass, pea-vine and corn-stalk, into the water furrow, with the hoes, early in the winter, and immediately casting a heavy furrow upon it." If the trash was too heavy to be turned under, the farmer could "have it dragged into piles with a heavy harrow, and litter the stock yards deeply with it; but by no means burn it." In this way the decaying residues would absorb nutrients from the manure and urine animals deposited in the stockyards. Cotton seed left over from ginning could also be "composted in some hollow or nook of bottom-land, where there is a deposit of rich earth." During dry weather in January the farmer should "haul out and distribute" these composts and manure onto the fields being plowed. In addition, ashes produced by clearing and burning woodlands were a valuable source of phosphorus and potassium. He exhorted farmers to "cut and roll logs; and, after burning them, carefully pile up the ashes and burnt earth, as an excellent manure for young corn—giving it a start over the grass, and being destructive to the cut worm." Affleck concluded that if nothing were "removed from the land, but the cotton fibre," if "the low lands [were] drained and protected from floods," if "the uplands [were] guarded from washing, by hill-side ditches or guard drains," if the planter practiced "occasional, even though defective, rotation of crops" and if "all of the cotton-stalks, leaves and seeds returned to the soil," soil exhaustion would be "so slight as to be scarcely perceptible."[21]

Affleck also understood the value of green manures. Peas were sometimes planted with the corn to provide additional forage or to help improve soil fertility by plowing the vines under. Though competition between the two crops could reduce the yield of corn, "so great is the valuable return of vegetable matter to the soil, by the plowing in of the corn-stalks and pea-vines; and so beneficial and ameliorating is the perfect shading of the soil, and so great the amount of feed for all kinds of stock, that it is questioned by a great many, whether the loss in corn be of any consideration." Two methods could be used to plant the two crops together. Affleck recommended that "half-a-dozen [peas] be dropped between each hill [of corn] at the second hoeing, that they may receive the benefit of at least one tending." The peas would then climb up the corn stalks, producing a good crop without reducing corn yields. On the other hand, "planting on thin or light sandy soils, alternate rows of corn and peas, the rows being of the usual distance apart, so much more light and air are admitted to both crops, that a larger yield may be obtained from each, with a much heavier cover of pea-vine, than if the pea were planted later and imperfectly tended. They, at the same time, shade the ground by the time the corn begins to tassel, protecting and fostering, instead of robbing, the corn-roots."[22]

In addition to using animal and green manures, farmers could purchase fertilizers from dealers such as Alfred Kearney on Magazine Street in New Orleans. Fine bone black cost eighteen dollars per ton, fine bone dust sold for thirty dollars per ton, "super-phosphate of lime" brought thirty-five dollars per ton, and "phosphate of Peruvian guano" cost forty dollars per ton. "Poudrette," dry night soil, could be purchased for two dollars per barrel.[23]

PROMOTING TEXAS AGRICULTURE

The leaders of many Texas communities tried to promote the advance of agriculture. As early as 1832 Stephen F. Austin offered prizes for the best cotton crops in his colony. In the 1840s merchants in Houston awarded silver cups for the first bale of cotton, the first five bales, and the first twenty-five bales brought to town. Agricultural societies were organized in 1843 in Lamar and Brazoria counties. The Red River Agricultural Society was established at Clarksville in 1851, and two years later it offered prizes for the best yields of cotton, corn, wheat, and oats, as well as for the best yearling mules, colts, calves, and hogs. The best locally made subsoil plow and two-horse turning plow also received prizes. Goliad advertised a fair in 1849 to exhibit livestock and equipment, and Corpus Christi held an exhibition in 1852, though livestock sales were "meager to what was expected," and the morals and conduct of those attending were criticized. The State Agricultural Society was organized in 1853 with planter Ashbel Smith as president, and counties were encouraged to form associated groups. Goals of the Agricultural Society of Brazoria County for 1855 included the following: "to introduce improved stock[;] to find by actual experiment the best agricultural implements to use—plows, hoes, ox yokes, axes, corn shellers, seed planters, and other things used in agriculture; to select and recommend the best system of governing, working, and attending to negroes; to improve the varieties of cotton, and to find out which kind is best adapted to our land and how the land should be cultivated to assure returns to the planter; to recommend the best kind of cane and the best method of cultivation; to introduce some system of which planters may avail themselves

> of the disposal of crops; to pay attention to the lesser crops of corn, potatoes, and the kitchen garden; and fruits and flower cultivation."[24]

Abigail Curlee's "A Study of Texas Slave Plantations, 1822–1865" and Elizabeth Silverthorne's *Plantation Life in Texas* are excellent sources of information about life on Texas plantations. Although it is not possible to describe the full range of plantation agriculture, brief descriptions of a few plantations will illustrate some of the diversity that existed.

GLENBLYTHE PLANTATION

In the 1840s Thomas Affleck established Southern Nurseries in Adams County, Mississippi, and wrote extensively on scientific agricultural practices. Recognizing the agricultural potential of Texas, between 1855 and 1860 Affleck gradually moved his family, possessions, and business to Gay Hill, northwest of Brenham. There he purchased thirty-four hundred acres for three dollars per acre and established a large, well-organized plantation named Glenblythe. In addition to developing and managing the plantation, Affleck established his Central Nurseries, bringing much of the stock from his Southern Nurseries in Mississippi but also importing materials from the North and from Europe. *Affleck's Southern Rural Almanac, and Plantation and Garden Calendar for 1860* carried a large number of advertisements for almost anything a planter might need. Affleck advertised that both his nurseries stocked a wide variety of fruit trees, grapes, strawberries, ornamental trees and shrubs, and flowers. They had stocks "sufficient to supply a moderate demand," and "none but Southern-grown trees, thrifty and well-grown, are ever sent out." Affleck advised the reader that "sales are made *only for cash,* or its equivalent." But he warned that "Uncle Sam's mail is not very trustworthy where cash remittances are concerned! When money is sent through the mail, the notes had best be cut in two, and mailed at different times, registering the letters." He indicated that "it is always safest to remit drafts on factors or others; or the orders may be sent through responsible business houses at any shipping point, as New Orleans, Mobile, Galveston, Houston, Lavaca, &c." In addition to selling plants, Affleck's Central Nurseries sold "pure and unadulterated mustang wine" that could be supplied in barrels or bottles. Touting the wine as "a pleasant and wholesome table drink," he recommended it as "a tonic for patients recovering from prostrating fevers, and for Females who may have been long in delicate health." In the same advertisement, Affleck's lumber and flouring mills offered to grind grain and could provide "lumber of all kinds, as wanted," including cedar, ash, and oak "dressed in any manner desired, as flooring and ceiling, dressed, tongued, and grooved." Doors, windows, blinds, and molding were also available. Affleck's almanac also made many detailed monthly recommendations about crop, livestock, garden, and plantation management. He emphasized the long view of plantation management and was a strong proponent of using deep tillage, plowing under plant residues, planting green manure crops, and applying animal manures to maintain soil productivity.[25]

After the Civil War Affleck apparently despaired of continuing to operate the plantation. In 1865 he advertised in the *New York Herald* that the plantation could be rented for one or "certainly not for more than two" years. A separate description, probably meant to be sent to prospective renters, provides our most detailed description of Glenblythe. The plantation consisted of thirty-five hundred acres, including sixteen hundred acres of timber and "the balance high, rolling, rich prairie." About nine hundred acres had been cropped for between two and five years, sixty to seventy acres of "sandy post-oak land" had been cropped for three years, about three hundred acres of "creek bottom, cedar-brake land" had been cleared and were "ready for the plow," and one hundred acres were an "excellent natural meadow" used to cut native grass hay. Livestock included "twenty head of No. 1 Kentucky mules," twelve to sixteen yokes of "well-broke" oxen, forty milk cows, three hundred "fine Sheep," a "large & excellent stock of Hogs," poultry, brood mares, a stallion, and "other horse stock." The beef cattle may well have been confiscated or sold during the war. Equipment included two mule- and three oxen-drawn wagons, a "small spring wagon," two "one-horse carts," "horse-scoop," plows, harrows, a roller, one "two-horse mower," "two of McCormick's Combined four-horse Reapers-&-Mowers," a "Portable Hay-press," and "Revolving rakes."

The plantation had a three-story building housing a flour mill, gristmill, cotton gin stand, and sawmill. They were powered by a recently repaired steam engine "with tubular boiler supplying abundance of steam." The engine could power the cotton gin stand, flour mill, and gristmill simultaneously, but the sawmill required so much power that it had to be operated alone. The flour mill had thirty-inch burr stones and could grind up to 70 bushels a day, which could be sold for five to six dollars per hundred pounds. The gristmill could grind 120 bushels a day. Affleck noted that the mills could be used to grind corn and wheat for neighbors, customarily keeping a toll of one-fourth of the cornmeal and one-fifth of the wheat flour. The wheat bran, which few customers wanted, could be fed to the plantation's hogs. Timber could be sawn for a toll of half the lumber. The plantation also had "a good Blacksmith's Shop, with two forges & full sets of tools," a frame "Cooper's Shop," and a "large Sugar or Sorghum mill" with three iron rollers to crush the cane and four kettles to boil down the juice. Affleck noted that he made "a good deal of syrup" that he sold and "traded for beeves, pork-hogs, bacon &c., at $1. per gallon."

The plantation house was "one of the most comfortable & commodious in the state; & situated in a very beautiful & elevated prairie valley, studded over with Live-oak & other groves." It was in "one of the most healthy" locations "on this continent" and was "well watered, by springs & large & permanent artificial tanks." The house had "six bedrooms, large & airy; dining-room & parlour; two large & pleasant halls; dressing & bath-rooms; Kitchen, laundry-room, & store-room." It had "two galleries each 50 × 12 feet, & three large enclosed galleries." Under the house was "a fine cement cistern" that provided an "abundance of delicious, pure & cool water." Other buildings included a "large smoke & meat house with large gallery, cisterns &c.," a carriage house, a granary, stables, corncribs, servants' houses, a poultry yard, and a pigeonry. Within 150 yards of the plantation house were five or six houses for farmhands. About two miles away, in the "plantation quarter," were the mill, a two-room overseer's house, a church and hospital, storehouse, and twenty houses of former slave families.

Affleck indicated that the plantation had been established the year before the Civil War began and had housed 120 slaves. He asked for an annual rent of six thousand dollars with four thousand dollars paid in advance and two thousand dollars "in satisfactory drafts on New York or New Orleans, payable at the end of the year."[26]

BARRINGTON PLANTATION

Anson Jones's Barrington plantation in Washington County is a good example of a small plantation developed in the 1840s and 1850s. Anson Jones, then Texas secretary of state, moved his family to Washington County in 1842 when President Sam Houston moved the capital from Austin to the town of Washington. Jones was elected president in September 1844 and took office in December. He called the Convention of 1845 in Austin, where the delegates voted for annexation by the United States, and he continued to serve as president until annexation in

February 1846. Sue Winton Moss used primary sources to describe Barrington's development and management, which are summarized below.[27]

Jones's responsibilities as secretary of state and president did not halt development of Barrington. In January 1844 he purchased 1,107 acres about four miles southeast of Washington for six hundred dollars. He immediately contracted with a local craftsworker to build a frame house and two log structures in exchange for 200 acres of the land. In 1844 he purchased 1,404 rails for $10.48 and began to build his fences. In January 1845 the Jones family moved into the house, and they continued to make improvements to it over the next two years, probably including sealing the interior walls with boards. The house consisted of six rooms, four on the first floor and two upstairs under the peak of the roof, with a masonry fireplace at each end of the house. A central passage, or "hall," ran from the front door to the back of the house, much like the typical "dogtrot" cabin with two log rooms separated by a covered passage. In 1845 Jones also purchased 100 acres of "cedar brake land," which he probably used to cut logs and fence rails. After retiring to Barrington in 1846, he continued to use his slaves and hired workers to build fences to protect his fields. In 1846 and 1847 work continued on the house and associated log plantation buildings, including a kitchen, a smokehouse, an office, a barn and stable, slave quarters, a carriage house, at least two corncribs, a variety of rail fences, a cedar rail hog pen, and a pen for his chickens and pigeons. In 1847 Jones planted his first cotton crop and built a "cotton house" to store the crop until it could be transported to a gin. This log structure probably had a puncheon or board floor, and boards covered the walls to protect the crop from weather.

Anson Jones's Barrington was probably typical of many small Texas plantations. From 1847 to 1856 Jones produced between nine and thirty-five bales of cotton per year, usually receiving between six and eight and a half cents per pound. He hauled his cotton to nearby gins that charged from 8 to 12 percent to process the crop. Jones sold his crops to various merchants, including R. A. French in Washington County, Frederick Scranton and Co. in Houston, George Butler and Bro. in Galveston, John H. Brower & Co. in New York, and Mill, McDowell & Co. in New York. In the 1850s Jones depended on his agricultural profits as his sole source of income.

Records indicate that the garden and orchard were important sources of fruits and vegetables, including sweet and Irish potatoes, cucumbers, squashes, field peas, roasting ears, snap beans, beets, cabbages, mustard greens, melons, turnips, and peanuts. The plantation experimented with peach, plum, apricot, apple, pear, cherry, quince, and nectarine trees, with the peaches probably producing the best results. Wild pecans and dewberries were also harvested. In the late 1840s the plantation produced enough tobacco to sell, wheat that was used to make bread, and some sugarcane for home consumption. Livestock for home slaughter and use included cattle, hogs, turkeys, chickens, and pigeons. During the late 1840s and early 1850s Jones kept eighty to ninety cattle and a few horses and oxen on his plantation. Most of his annual agricultural income came from the sale of livestock and cotton.

Jones continued to buy and sell small amounts of land, and in 1851 Barrington consisted of about 695 acres, approximately half of them from the original 1,107-acre purchase. In 1855 he bought an additional 463 acres for three thousand dollars. After being rejected in his bid to become U.S. senator in 1857, he became increasingly moody, sold Barrington in December 1857, and in January 1858 committed suicide in Houston. The Jones house has been restored and moved to Washington-on-the-Brazos State Park, where it forms part of an exhibit of antebellum Texas life.

MONTE VERDI PLANTATION

In 1841–42 Julien Sidney Devereux, along with his father, John William Devereux, and approximately sixty-five slaves, moved from Macon County, Alabama, to Montgomery County, Texas. There they established Terrebonne plantation on the San Jacinto River southwest of the town of Montgomery. Dorman H. Winfrey's *Julien Sidney Devereux and His Monte Verdi Plantation* summarizes the history of Terrebonne and Devereux's second Texas plantation, Monte Verdi, in Rusk County.[28]

In 1843 Julien divorced his wife, Adaline, from whom he had separated in 1840 because of her "violent and passionate temper" and "habits of intemperance." Three months later he married Sarah Ann Landrum. Though Terrebonne was productive, both Julien and John became disillusioned with the heat and humidity of Montgomery County. John wrote that "I have been three years in this poison'd atmosphere—drank bad water and breath'd foul air—have borne northers and scorchers—inundations [and] droughts—pestilence endemic and epidemic—chill[e]d with cold & moisture and again panting and exhausted with hot sirocco air surcharged with malaria from stagnant lakes." In 1845–46 Julien sold Terrebonne and moved the family to Rusk County near the small community of Gourd Neck, where he bought the first parcel of what would eventually become a ten thousand–acre plantation. By March 1846 Devereux's slaves were building cabins and smokehouses and clearing and burning brush. Despite the Devereux's move to what they considered a healthier climate, during the summer and fall of 1846 the people of the entire region, including the Devereux family and their slaves, suffered from fevers. Concluding that their home was "too near the Angelina River," the family moved to a new house. In June 1847 John William Devereux, seventy-eight years of age, died of "chronic dysentary [*sic*] and dropsy," and Julien inherited eighteen of his twenty-nine slaves, with an average value of $275 each. In 1847 Devereux produced better corn and cotton (104 bales) crops than he had ever grown in Montgomery County. Then in Septeber 1848 his son died at the age of two and a half years. By 1849 Devereux had accumulated sixty-two slaves valued at $300 each, 3,027 acres of land valued at $1 per acre, 150 head of cattle worth $4 each, four hundred hogs valued at $1 each, twenty horses and mules worth a total of $1,200, and three wagons worth a total of $300. Over three-fourths of his total taxable property of $24,127 was in slaves, and he was poised to become one of the largest cotton producers in Texas. It was about this time that Devereux named his growing plantation Monte Verdi.

By 1850 Devereux was among the three largest cotton producers in east Texas, and his 120 bales placed him among only sixty-six Texian planters who produced more than 100 bales in that year. In 1850 a total of 7,747 Texians owned at least one slave, and Devereux's seventy-four slaves, forty males and thirty-four females, placed him among the ninety-two slave owners who owned more than fifty. By 1853 the number of his slaves had increased to eighty-four. In an application for insurance in the same year, Devereux declared that he had two gin houses and a Burow improved corn mill. One of the gin houses was forty-one feet square with a fifty-saw improved cast-iron Pratt gin; the other was thirty-six feet square with a forty-five-saw Pratt gin. The cotton press had a walnut screw. The gins and press, turned with mules, had a total value of $2,500. The U.S. Census of Agriculture for 1850 indicates that Devereux owned 420 acres of improved land and 3,892 acres of unimproved land. The cash value of the farm was $16,000, including farm implements and machinery worth $2,000. His livestock was valued at $2,500 and consisted of four horses, twenty-one asses and mules, forty milk cows, sixteen working oxen, one hundred other cattle, twenty-five sheep, and three hundred swine. His produce for the year included three thousand bushels of corn, one hundred bushels of oats, 120 bales of cotton (each weighing four hundred pounds), one hundred pounds of wool, two hundred bushels of peas and beans, one thousand bushels of sweet potatoes, and five hundred pounds of butter. He slaughtered animals valued at $360.

Winter, usually December and January, was the season to kill hogs and put up pork for later uses. By the end of January 1849 sixty-one hogs had been killed at Monte Verdi, and 10,444 pounds of pork were put up. In 1850, 1853, and 1854 the totals were 12,059, 13,695, and 14,793 pounds, respectively. In 1855 Devereux sold 132 cattle for a total of $903.

Devereux ginned cotton and ground corn for the public, and in 1848 he put his customers on notice that they should "bring bagging, rope and twine with the first load or it cannot be received," and "all cotton will be considered delivered as soon as it is packed." Devereux also announced that he would "not be liable for accidents" at the gin or the gristmill, and he would not be responsible for "sacks or bags of corn or meal unless the sacks are plainly marked."

The cost of maintaining Monte Verdi, with its large population of slaves, was substantial, and Devereux paid large bills to merchants, including $3,123.48 to McClarty & Son in Henderson in June 1852 and $3,552.34 to Charles Vinzent at Mount Pleasant in August 1854. But Devereux, like almost all planters, kept meager financial records. He bought on credit and paid when the income from his crops and livestock permitted. Like other planters, he depended on statements from factors and other merchants to determine what he had spent and how much he owed or how much credit he had accumulated. Under some conditions merchants even asked him for advances, as when Charles Vinzent requested $1,000, to be repaid within two or three weeks, to help some friends who were in financial difficulty as a result of a decline in cotton prices in New Orleans. As a buffer against unexpected financial needs, Devereux often kept large amounts of cash on hand; in May 1853 he noted that he had $1,571 in bank bills, gold, and silver.

From 1846 until 1852 Devereux enjoyed the services of two excellent overseers, William Howerton and A. C. Heard, often assisted by a devoted slave named Scott. Devereux praised Heard for his willingness to work in the field with the slaves in cold and rainy weather, writing in March 1846: "This morning raining—hands continued planting through the day in drizzles of rain—light rains don[']t drive Heard out." Howerton earned between $300 and $350 per year plus twenty bushels of corn from 1847 until 1852. In 1850 Devereux, possibly anticipating Howerton's departure, began to look for another overseer. He wrote a friend that he had "about 25 good hands[,] all family negroes, and easily governed[.] I am contented with having moderate work done & would not have a severe or cruel overseer on any terms." But his next overseer, a man named McKnight, was "a worthless, drunken, stubborn[,] ill natured fellow." In 1855 Devereux fired McKnight because Monte Verdi had "made scarcely any cotton crops for the last two or three years." He wrote that "if McKnight had continued a year or two longer my plantation would have been broken up & we should have been compelled to sell our home or some of the negroes."

Devereux felt that slavery was justified, citing repeated references to it in the Bible. He wrote that "slavery has been tolerated among the Jews and persons represented as *God's people*—how do the Fanatics urge that slavery is forbidden by the scriptures . . . [?]" He clearly felt that he treated his slaves well, allowing them to miss work when they were sick or the weather was bad. He also permitted at least twelve of his slaves to cultivate and sell cotton, corn, and vegetables and manage the money they received. In 1850 the total value of their crops was $737.05, and in 1853 Charles Vinzent of Mount Enterprise bought the slaves' cotton crop for $300. However, in 1846 Devereux approved when his overseer Heard gave three of his slaves "a good thrashing which they well deserved for impudence." In 1850 he purchased a "clog made to put on negro man Ben." The fifteen-pound weight was presumably designed to discourage Ben from running away. In 1852 he paid a man named Birdwell ten dollars for returning a runaway slave, "the first occurrence of the kind" that had ever happened to Devereux.

The year 1855 was the high point of Devereux's career as a planter and east Texas leader. His total taxable property that year was worth $50,307. It consisted of seventy-five slaves valued at $30,000; 8,253.5 acres of land worth $16,507; thirty horses and mules valued at $1,800; three hundred head of cattle worth $1,500; and miscellaneous property worth $500. Devereux began construction of the large Monte Verdi plantation house and was then elected to the state legislature representing the Democratic Party. He traveled to Austin for the legislative session that began in early November but returned due to poor health shortly before the session ended in February 1856. His health continued to decline due to what he termed an "enlarged spleen," and on May 1, 1856, at the age of fifty, he died at Monte Verdi.

Devereux had become one of the wealthiest men in east Texas. The inventory of his estate after his death included 10,721 acres of land, eighty slaves, six horses, fifteen mules, fifteen yokes of oxen, three hundred cattle, three hundred hogs, 230 sheep, sixteen goats, four road wagons, one two-horse wagon, one pleasure carriage, two gin stands, one horse mill, one set of blacksmith tools, thirty-eight plows, twenty-five weeding hoes, fourteen grubbing hoes, eighteen axes, two broad axes, twelve log chains, eight iron wedges, four bedsteads, six feather beds, six cotton mattresses, twelve chairs, one bureau, one clothes press, one cupboard, one lot of books, one double-barreled shotgun, two rifles, four pistols (probably one derringer, two single-shot, and one repeater), six baking ovens, two wash pots, two cooking pots, one brass kettle, and $1,912.40 cash. Sarah continued to manage and grow Monte Verdi, completing the grand Monte Verdi plantation house in about 1857 and rehiring William Howerton as overseer in 1858. The U.S. Census of 1860 valued her personal estate at $66,000.

PEACH POINT PLANTATION

In 1830 Stephen F. Austin was able to secure a grant from the Mexican government of eleven leagues of land (one league equals 4,428 acres) for his brother-in-law James F. Perry and sister Emily Margarita Bryan Perry. The land was "to be selected on any vacant lands in Austin's colony" on the condition that the family settle the land by January 1, 1832. Austin advised Perry that "there is a fine opportunity here for a good Merchant, and a regular trading schooner to ship produce . . . to Tampico and Vera Cruz would make money rapidly." Commodities available for export could include corn, cotton, sugar, beef tallow, pork, lard, and mules. Austin suggested that the family move in the fall of the year in order to avoid illness and bring "breeds of English cattle, [because] nature never made a better place for stock." In addition, they should bring seeds, bedding, and furniture. Austin reminded Perry to indenture his slaves before a judge or clerk before he left the United States because slavery was illegal, though indentured servants were not. He advised that the eleven leagues would cost Perry "about $1,000 including everything, and you will be allowed 4, 5, and 6 years to pay a part of that in and the balance can be settled by me easily."

Using primary sources, including family letters and plantation record books, Abigail Curlee and Marie Beth Jones have described the development and management of Peach Point plantation by several generations of the Perry family.[29]

Following Stephen F. Austin's advice, James Perry and his family arrived in San Felipe in August 1831. The party consisted of James; his wife, Emily; her four children by James Bryan, who had died in 1822 (William Joel, Moses Austin, Guy, and Mary Elizabeth Bryan); the couple's two children (Stephen and Eliza Perry); and James's niece Lavinia. In the winter of 1831 or spring of 1832 the family moved to

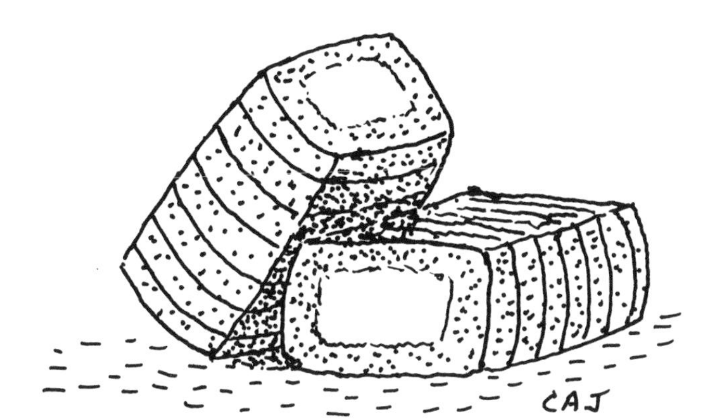

Bales of cotton

Chocolate Bayou in present-day Brazoria County, where Perry established a ranch. In December 1832 they moved to Peach Point, about ten miles below Brazoria. The entire plantation consisted of five leagues on Chocolate Bayou and five other properties of one or two leagues each, for a total of twelve leagues. Of the total, about 15 percent was farmland; the remainder was grazing land.

Like many others, the family and its slaves suffered from cholera and malaria during the epidemics of 1833. Emily nearly died of malaria; and in August Mary Elizabeth, Emily's eleven-year-old daughter by her first marriage, died of cholera. In late October Perry wrote to Austin that the family had not been "entirely clear of sickness since June." The crops planted early in the spring did well, and the family produced around nine hundred bushels of corn and "plenty of pumpkins." But their cotton could not be planted until the end of June, and the boles did not open before the crop was injured by an early frost. This was particularly disheartening because cotton was bringing from sixteen to eighteen cents a pound in New Orleans.

In 1834 Perry's cotton crop was small because he planted only a small amount and one-third of that was destroyed by worms. By May 1835 Perry had expanded his cotton land to sixty-five or seventy acres, but there is no record of how much was produced. In 1836, because of the Texas Revolution, Perry produced only twenty-two bales of cotton, and food crops were also short. There is no record of the crop in 1837, but Perry borrowed five hundred dollars from his factor in New Orleans, indicating that conditions were not particularly good. Beginning in 1838 there are relatively good records of the cotton crops. The number of bales, number of hands who participated in picking, weight per bale, and net return after expenses are given below.

Year	*Cotton (bales)*	*Hands*	*Weight/Bale (pounds)*	*Net Return ($)*
1838	127	14	—	—
1839	106	20	—	3,744
1840	103	25	—	4,562
1841	89	19	—	4,339
1842	50	19	—	1,306
1843	61	20	481	2,701
1844	118	28	—	3,133
1845	130	22	—	4,645
1846	10	—	—	460
1847	105	27	526	1,217
1848	82	23	513	—
1849	39	19	467	—

According to the census of 1850, in 1849 Perry had forty-three slaves. If the ratio of field hands to the total population of slaves remained relatively constant, during the 1840s Perry probably owned about forty to fifty slaves.

The best record of plantation work during this period when the plantation concentrated on cotton production comes from Stephen Perry's journal for 1848, which provides details of daily work from the middle of January until the end of June. From the beginning of 1848 until February 6 the plantation gin finished the last cotton from the 1847 crop. During the last half of January the slaves were kept busy making and weighing cotton bales, rebuilding fences, cleaning gutters, shelling corn, splitting rails, and building a chimney for one of the slave cabins. Between January 27 and February 1 forty-seven hogs were killed, and the meat was salted. During the first week of February, forty bushels of corn and twenty-two bales of cotton were carried to Aycock's landing for shipment. On February 8 the hands began cleaning the cornfields and pulling and rolling cotton stalks in preparation for plowing, which began February 9. Three plows were in operation on February 10, and plowing and preparing other fields for plowing continued throughout the month and into March. Corn planting began on February 17 and finished on February 25. As soon as the corn began to emerge, the slaves began "minding birds" intent on eating the germinating grains. On February 26 the hands began "throwing up cotton ridges" in preparation for planting. Cotton planting began on March 1, but the slaves continued to "make cotton ridges" until March 15, when the crop had all been planted. Other activities during the first half of March included plowing the sweet potato field, planting the potatoes, cleaning up sugarcane fields, and finishing baling all the cotton from the previous crop. During the last half of March the hands cleaned out drainage ditches, built fences, and began weeding the corn with light plows and hoes. Stephen noted in his journal that they did not harrow the corn this year, and "I do not think we did right." The sweet potatoes began to emergeon March 26.

Stephen remarked that "the months of January[,] February[,] and March have been exceedingly favorable to the Planters." But he complained that "the atmosphere has become impure which has produced sickness among the negroes." The chief complaints were "pains in the breast and sides." From January 17 until April 1 thirteen field hands were sick a total of sixty-seven days—one field hand named Allen missed twenty-four days.

On March 31 and April 1 the hands finished the first weeding of the corn and began weeding the cotton with light plows and hoes. As in the cornfields, weeding the cotton proceeded from one field to the next until it was completed on April 17. On April 15 the hands began the first weeding of what was apparently a small area of sugarcane. On April 18 they "commenced ploughing corn[,] hilling it up[,] and ploughing out the middles. Corn looks very well indeed[;] wants rain very much[;] good stand in all of it except the cotten [*sic*] ground[;] replant not all come up." On April 19 seven plows were in use weeding the corn for the second time, and the plowing and hoeing continued until April 27. A small rain on April 26 and a hard rain the night of April 28 relieved the drought. Because the fields were wet, eight hands set out sweet potato sprouts, one cooked, two hauled wood, two hands ground corn and drove up cows and calves, and three worked on the road.

Field work began again on May 1, with four plows starting to weed the cotton for the second time, and twelve hoe hands "finished set[t]ing out potatoes before bre[a]kfast and went to hoeing cotten [*sic*]." On May 2 eleven hands finished replanting cotton in one field and began hoeing it in another. Cotton occupied all hands on May 3 and 4, using shovel plows and a "one horse plow." In addition, hands were "cutting the Cotten [*sic*] out to a stand." Stephen noted that the corn was tasseling and they had "fine prospects for a good crop of cotten [*sic*]." On May 10 hands began "ploughing out the middles with the sweeps that

Cut next to the Corn," and on May 11 the second weeding of the cotton was complete. Weeding the corn continued until May 22, when it appears to have been "laid by." At least some of the sweet potatoes were planted in the cornfield, and on May 22–24 hands were weeding potatoes and "making potatoe [*sic*] ridges in the corn." The hands continued weeding the cotton with plows and hoes during the last week of May and first week of June. A slave named Tom ran away on May 30 but returned on June 7. The sweet potatoes and cane were also weeded on June 7 and 8. Rain on June 8–9 slowed the field work, but the hands shucked and shelled corn and cut wood. It was dry enough to hoe cane for part of June 9 and to begin plowing after dinner on June 10. It rained again on June 14, and hands were employed with "some getting board timber and some ho[e]ing down the large weeds in the fields." Some "cleaned the ridge where the Bo dark [bois d'arc, or Osage orange trees used for hedges] is planted," and others spent part of the day bringing wood to make cotton baskets. Hoeing the cotton began again on June 15, and plowing began the next day. On June 17 Stephen notes that a slave named "Wesly has a sore shine [shin] which is nearly well, we have been doctering [*sic*] him by applying a plaster of fresh cow menure [*sic*] which is very good." On June 20 sweeps were added to the plows and hoes weeding the cotton. On June 27 the hoe hands finished the cotton and began hoeing the potatoes planted in the corn and "cutting down weeds in the corn." Stephen Perry's daily journal entries stop at the end of June, though he notes a few important milestones later in the year. For example, cotton picking began on August 17 and ended on October 11. Heavy frosts killed the cane on November 4 and 5. Digging sweet potatoes ended on November 21, and ginning began on November 24.

We know that Peach Point began producing commercial crops of sugar in 1844 because a letter from James to W. Hendley & Co. in September 1845 complains of a lack of details in the company's charges for transporting the plantation's sugar and molasses to Galveston and selling them in New York. However, there is little other information during these early years. In 1848 there is a record of cane being harvested after a severe frost on November 4, though it is clear that much more labor was expended on the corn and cotton crops. In 1849 and 1850 Perry built a sugar house; bought a sugar mill, complete with steam engine and kettles, for $7,337.66; purchased $57.20 worth of "seed" cane; and bought molasses barrels and sugar hogsheads to store and transport the crop. Sugar making began on December 12, 1850, after a freeze killed the cane, and it began to sour. The crop produced 45 hogsheads of sugar and 165 barrels of molasses. The crop grown in 1851 produced 202 barrels of molasses and 60 hogsheads of sugar, each probably weighing from eleven hundred to eighteen hundred pounds. Expenses included $25 for work on the sugar house furnace and $676 for the purchase of 321 barrels and 151 hogsheads prior to harvesting the crop.

The sugar crop grown in 1852 suffered a number of setbacks. The furnace and a pump in the sugar house had to be repaired before the crop could be harvested. The corn house and stables burned in November, destroying over three thousand bushels of corn and most of the plantation plows, harrows, and carts. Though the cane harvest began in October, a freeze in early January effectively ended the harvest. Despite these problems, a warehouse bill indicates that Peach Point produced at least 450 barrels of molasses and 71 hogsheads of sugar. Costs included about $650 for making and repairing barrels for the 1852 and 1853 crops and $420 to pay Jesse Munson, who served as engineer in the sugar house. In addition, Perry often hired ten or fifteen additional slaves from surrounding plantations during the cane harvest. He typically paid $20 to $30 per month for each one.

Records for the 1853 sugar crop are poor, perhaps because in September James Perry, his son Henry, and his daughter Eliza contracted yellow fever in Biloxi, Mississippi. James had accompanied Eliza to Biloxi where she was undergoing "water cure" treatments for a chronic illness, and Henry had come to begin his study of the law. But in July and August a yellow fever epidemic began, making its way from New Orleans to Biloxi. James and Henry died, and only the intercession of Eliza's doctor prevented their burial in a mass grave.

Stephen Perry inherited the plantation, and in 1854 the sugar crop was good. Processed by December 27, it yielded a net profit of $5,365 after deducting freight, wharfage, auction, cooperage, interest, commission, and insurance costs. In 1855 the net profit (after $2,579 in expenses) rose to $6,781 on 180 barrels of molasses and 140 hogsheads of sugar marketed in Galveston, Baltimore, and New York. Prices of sugar from 1851 to 1860 ranged from five to six cents per pound. P. A. Champomier's annual "Statement of the Sugar Crop Made in Louisiana" shows that Peach Point produced 220,000 pounds of sugar in 1854, 214,000 pounds in 1855, and 240,000 pounds in 1858. This made Stephen Perry the ninth-largest sugar producer in Texas in 1858.

In addition to growing the main cash crops, cotton and sugarcane, Peach Point plantation supplied its own needs for corn, sweet and Irish potatoes, tobacco, eggs, chickens, pork, lard, Muscovy ducks, turkeys, geese, butter, pecans, tallow, hominy, and soap. Records indicate that the plantation produced between eighteen hundred and thirty-eight hundred bushels of corn in 1841, 1846, 1847, and 1848. It consumed most of the crop, selling only small amounts of corn in the shuck and as grain, fodder, and meal. In 1845 two hundred bushels of sweet potatoes and eighteen bushels of Irish potatoes were sold. In the last few days of January 1848, forty-seven hogs were butchered, probably only a fraction of those killed throughout that winter. In 1843, 1844, and 1845 an average of 300 pounds of pickled pork was sold at eight cents a pound. In 1848 the plantation sold twelve hogs, 1,744 pounds of bacon at eight cents a pound, and 334 pounds of lard at nine cents a pound. Sales records from the Chocolate Bayou ranch are scarce. Records indicate that the ranch produced an average of about thirteen hides per year with a value of less than two dollars each. Between 1839 and 1846 recorded cattle sales ranged from $40 to $240 per year; however, between 1856 and February 1859 N. S. Davis was overseer, and he sold 238 cattle for a total of $5,532, a net return of $4,224.

When Perry came to Texas in 1831, he brought his slaves with him, though we do not know how many. There are good records of the number of field hands used to harvest the cotton crop but almost no record of the number of domestic and other slaves on the plantation. There are also relatively few records of purchases or sales of slaves. We do know, however, that Perry's stepson, William Joel Bryan, purchased several slaves. In 1840 he married Lavinia Perry, and the couple was given Durazno plantation, an extension of Peach Point. In 1842 Bryan paid one thousand dollars to a man named Hopkins for "negress Ann and child"; three thousand dollars to a doctor named Smith for "negress Tamar, negroes Donor & George"; and one thousand dollars to Emily M. Perry for a boy named Frank. Slaves were sometimes allowed to farm small plots of land, about one acre each, and to raise hogs. Records indicate that between 1839 and 1841 several slaves earned between thirty and fifty-five dollars each per year for cotton that they grew and picked. In addition, slaves were allowed to raise and sell small amounts of corn and bacon, and they received one dollar for each Sunday that they worked.

Beginning with William Joel Bryan in 1836 and 1837, Peach Point had a total of eighteen overseers between 1836 and 1858. Though several worked as long as two years, others lasted only a few days, and their salaries ranged from about $240 to $600 per year. Peach Point also employed a number of free artisans for jobs such as doing carpentry and shopwork, installing the sugar mill, making barrels and hogsheads, and managing the sugar house.

Rutherford B. Hayes, Guy Bryan's classmate at Kenyon College in Ohio, visited Peach Point in January and February 1848. Among his observations were that "white and black mechanics all work together" and the white workers were "generally dissolute and intemperate." But he was impressed with the wildlife. On a trip to the mouth of the Bernard River "to fish and eat oysters," he observed deer, cattle, cranes, wild geese, brant, ducks, plover, prairie hens, and the Lord knows what else, often in sight at the same time." Letters also mention killing bears, turkeys, ducks, deer, wild hogs, "and all kinds of wild game." Hayes observed: "These Texans are essentially carnivorous. Pork Ribs, pigs' feet, veal beef (grand), chickens, venison, and dried meat [are] frequently seen on the table at once." Other dishes included cornbread, several varieties of beans, sweet potatoes, turnips and their greens, coleslaw, hominy, preserved fruit, butter, sweet milk, buttermilk, and coffee. An assortment of jellies, jams, preserves, pickles, relishes, and cakes made of wheat flour complemented the meals. The slaves consumed the same cornbread and vegetables, which were produced on the plantation, but their meat was mostly salt pork, rather than the variety enjoyed by the family. The future president was also impressed by Peach Point's house, described as "beautifully situated on the edge of the timber, looking out upon a prairie on the south, extending five or eight miles to the Gulf, with a large and beautiful flower garden in front." The social life included "no end of entertainment—balls and parties rapidly followed one another, the guests riding ten, fifteen, and even twenty miles, arriving early in the afternoon, and remaining for nearly twenty-four hours, the great plantation house supplying room for all."

SLAVERY

As the number of slaves increased in the 1840s and 1850s, well-organized slave markets developed. Slaves could be purchased at permanent markets in New Orleans, Shreveport, Houston, and Galveston, with the cost, like that of other goods, charged by the planter's agent, or factor, to his account. For example, the *Texas Almanac for 1860* advertised J. S. Sydnor's "auction and commission business" on the Strand in Galveston. Sydnor noted that "within the last several years, the number of Negroes consigned to me for sale, has induced me to give the business more particular attention." As a result, he reduced his commission on the sale of slaves to 2.5 percent and arranged to employ those offered for sale "directly adjoining my store which will at least pay their board, and they will always be on hand when parties desire to see them." In addition, because servants were "in great demand for hire about the city," Sydnor offered to help their owners "hire them out, and produce a revenue, instead of accruing a bill of charges." Dealers would also bring slaves to county seats far from the permanent slave markets for sale to planters in the region. In all cases, sales resembled those of livestock, with the buyer carefully examining the slave's teeth and physical condition. Though some slave owners tried to keep slave families together, purchasing family members they did not need, in many cases children were torn from their parents, and couples were separated, never to see one another again.[30]

In general, young and vigorous men and women, especially those with useful skills, brought the highest prices. This is illustrated by the sale of slaves from the plantation of Alexander Moore, who owned a six hundred–acre plantation on Peach Creek in Wharton County. When he died in 1853, thirty-one slaves were sold. Three couples were sold together with the highest prices paid for the youngest: Jobe (25 years) and Martha (20)—$3,600; Phil (25) and Ritta (28)—$2,600; and Prince (60) and Elsey (50)—$1,000. Five women were sold with their children: Minerva (31) with Isaiah (8), Austin (6), Esau (5), and Lizzy (1)—$3,200; Eliza (28) with Jordan (5), Israel (4), and Angeline (1)—$2,800; Charlotte (35) with Jim (4) and Caroline (1)—$1,500; Margaret (30) with John (2)—$1,525; and Isabel and her infant—$1,700. It is unclear who the fathers of these children were, but if they were slaves, they were not sold with the mothers and children. Six men and boys were sold separately: Major (31)—$2,000; Simon (28)—$1,450; Calvin (24)—$2,000; Jeff (19)—$1,750; Elisha (15)—$1,700; Wash (14)—$875; and Henry (13)—$900. Two young women, Maria (26) and Harriet (15), were bought for $1,600 and $1,100, respectively.[31]

Though most slaves worked for their masters, owners frequently hired out some of their slaves. For example, in 1823 Stephen F. Austin hired three slaves, two men and a woman, from Jared Groce. The period was one year at a cost of $18 a month for the woman and $15 a month for each man. A boy was furnished to care for the woman's child. Groce was responsible for the slaves' clothing, and Austin promised to treat them well. It was agreed that if the slaves ran away or died, Groce would bear the loss; time lost due to sickness would be Austin's loss. Olmsted reported that in 1854 "ordinary" slaves could be hired for $150 to $250 per year; "first rate hands" brought $300. At a "'hiring' of Negroes" at the county courthouse in Caldwell, "eight or ten were hired out at from $175 to $250 per annum—the hirer contracting to feed them well and to provide two substantial suits of clothing and shoes." In 1853 the *Texas State Gazette* reported that nine slaves had been hired at an average of $280 per year. Slaves were also allowed to hire themselves out, usually for no more than one day a week.[32]

As in other parts of the South, a social hierarchy existed among slaves. At the top were the house servants, those who cooked, cleaned, mended, cared for children, and otherwise served the master and his immediate family. Below them were the craftsworkers, including carpenters, brick makers, masons, weavers, seamstresses, tanners, cobblers, gin operators, sugar makers, gardeners, and stockmen. Some slaves, such as Henry Harris, who lived on the Bolton plantation in Wharton County from 1836 to emancipation, mastered many crafts. He built houses, tanned hides, made shoes for the slaves, twisted rope from cotton and horsehair, made coffins, wove oak and willow baskets, made yokes for oxen, and molded bricks. At the bottom of the hierarchy were the field hands, who plowed, planted, weeded, harvested the crops, and did other related work on the plantation.[33]

With the exception of house servants, slaves were usually managed by an overseer. Responsible for keeping the plantation running smoothly and profitably, the overseer lived on the plantation and was usually paid $150 to $600 per year, sometimes receiving a percentage of the crop and livestock production. Regardless, the overseer had the difficult job of pleasing the master, producing a profit, and keeping the slaves, usually the planter's major source of wealth, healthy and productive.[34]

Plantations varied in the kinds and quality of food and clothing they provided their slaves. Some planters hoped to protect their investment by feeding their slaves well. Others were far less generous. Regardless, most planters tried to minimize the costs of food, using their workers to produce corn, potatoes, sweet potatoes, southern peas, and a variety of other vegetables in large plantation gardens. In many cases, they also allowed slaves to produce food in their own small family gardens. Beef and pork were supplemented with wild game, including deer, raccoon, opossums, rabbits, and turkeys, all of which were sometimes hunted by slaves. Slave children, cared for by older adults while their parents were at work, were usually fed from a large common bowl, tray, or trough around which they all gathered to eat with spoons. When slaves prepared meals in their own cabins, they were normally issued provisions once a week. Typical rations for a working hand were two and a half to three and a half pounds of bacon per week, which could be reduced if milk, butter, and molasses or syrup were supplied. If dried beef was the source of meat, five or six pounds were a typical weekly ration. A peck of cornmeal or a supply of sweet potatoes provided needed carbohydrates. Rations might also include fresh beef or pork, tobacco, vegetables, and even wheat flour. However, some planters felt that issuing weekly rations resulted in wasted food and poor nourishment. As a result, some fed their slaves from a common kitchen. At Bernardo plantation cooks began preparing the food at four o'clock in the morning. When the morning gong sounded at daybreak, the slaves came to the dining hall adjacent to the kitchen for a large cup of strong coffee. They then went to the field and worked until a breakfast of biscuits and meat was brought in buckets at seven o'clock. Dinner was brought to the field at noon, and a hot supper was served in the dining hall at six o'clock in the evening.[35]

In order to maintain a productive and growing workforce, most owners insisted that slaves marry and begin producing children as soon as possible. If women resisted, they were sometimes forced to cohabit with men, including the owner or overseer. Older slave women and girls usually cared for the smaller children, allowing their mothers to return to the field or to other jobs. In addition, as soon as slave children were old enough, they were given small chores. By the time they were ten, they were often working in the fields or were apprenticed to learn a craft.[36]

Not infrequently slaves ran away from their owners, often attempting to make their way to Mexico, where they were welcome and free. In order to recover their valuable property and discourage others, planters aggressively pursued runaways, sometimes even killing runaway slaves. Noah Smithwick reported that Jim, one of the Pleasant McNeel family's slaves, "openly announced his determination to leave, and acting on the impulse, threw down his hoe and started away." McNeel ordered him to return to work, and when Jim kept walking, raised his rifle and threatened to shoot him. Smithwick continued, "Jim, however, kept on, and true to his threat McNeel shot him dead." When runaway slaves could not be tracked and caught with packs of dogs, owners often placed advertisements in newspapers. From the late 1830s through the Civil War, Texas newspapers notified their readers of runaway slaves, often announcing rewards of from twenty-five to two hundred dollars.[37]

Texas planters and their overseers varied widely in their treatment of slaves, from relative kindness to great cruelty. Some did not allow their slaves to be whipped and let them make their own money by hiring themselves out and growing and selling their own crops. Others were extremely cruel, making slaves work long hours without adequate food or clothes and beating them mercilessly. Most probably fell between the extremes, attempting to protect their investment by feeding and clothing their slaves adequately, yet willing to punish, even whip, them to maintain discipline. This chapter is far too short for an adequate discussion of that "peculiar institution," slavery. In addition, it is difficult for anyone living today to appreciate the cruelty and inhumanity of slavery in the antebellum South. However, it would be inexcusable not to give the reader some concrete examples of both its variety and its brutality. Following are summaries of four firsthand accounts from the Federal Writers' Project in the 1930s, when most of the former slaves interviewed were well into their eighties. Some readers may be offended by the dialect and language used by the project. Some may question the accuracy of their stories. But these narratives provide a sense of how some Texas slaves recalled their past, as well as the corrosive effects slavery had on their lives.

Rose Williams, born in the 1840s on William Black's plantation in Bell County, recalled that "Massa Black am awful cruel, and he whip the colored folks and work 'em hard and feed 'em poorly. We-uns have for rations the corn meal and milk and 'lasses and some beans and peas and meat once a week. We-uns have to work in the field every day from daylight till dark, and on Sunday we-uns do us washing. Church? Shucks, we-uns don't know what that means." But Black was a slave trader, and Williams recalled, "He buy and sell all the time." As a result, at about the time the Civil War began, she, her mother, and her father were sold with a group of about ten slaves. When they got to the sale, "lots of white folks . . . come to look us over." A man named Hawkins began to talk with Williams's father, who told him, "Them my woman and childs. Please buy all of us and have mercy on we-uns." Williams recalled that Hawkins told her father, "That gal am a likely-looking nigger; she am portly and strong. But three am more than I wants, I guesses." However, Hawkins purchased all three members of the family, outbidding two other men to pay $525 for her. He took them to his "nice plantation" where he had about fifty adult slaves "and lots of children." The first thing that Hawkins did was give them a cabin and "plenty meat and tea and coffee and white flour." Williams had never before tasted white wheat flour. The plantation's twelve slave cabins were made of logs and had "a table and some benches and bunks for sleeping and a fireplace for cooking and the heat." There was no floor, "just the ground." Williams recalled that Hawkins was good to his slaves and did not force them to work too hard. He allowed them to fish and "have reasonable parties," but they were never taken to church or allowed an education. Even so, Williams felt that there was as much difference between Hawkins and Black "as 'twixt the Lord and the devil." Though Rosa Williams and her family were much better treated by Hawins than Black, she was forced to live with a slave named Rufus because, as she recalled her mistress saying, "You am the portly gal, and Rufus am the portly man. The massa wants you-uns for to bring forth portly children." And she recalled that the next day Hawkins himself told her, "Woman, I's pay big money for you, and I's done that for the cause I wants you to raise me childrens. I's put you to live with Rufus for that purpose. Now, if you doesn't want whipping at the stake, you do what I wants." Williams, of course, had little choice. She wrote, "I thinks 'bout Massa buying me offen the block and saving me from being separated from my folks and 'bout being whipped at the stake. There it am. What am I's to do? So I 'cides to do as the massa wish, and so I yields."[38]

James Green was the free son of a slave woman belonging to John Williams in Petersburg, Virginia. He gained his freedom because his father, "a full-blood Indian," had "done some big favor for a big man high up in the courts, and he gets me set free." However, when Green was twelve years old, Williams took him to town and sold him for eight hundred dollars to John Pinchback, who immediately took Green to his plantation near Columbus, Texas. Probably suspecting that Green would run away as soon as he had the opportunity, the day after they arrived at the plantation Pinchback chose him to train the dogs used to track runaway slaves. "I's told I had to play I running away and to run five mile in any way and then climb a tree. One of the niggers tells me kind of nice to climb as high in that tree as I could if I didn't want my body tore off my legs." The experience of being tracked and treed by the hounds, even if it was only practice for the dogs, was enough for Green, and he recalled that "after that I never got thinking of running away." Green reported that Pinchback "breeds niggers as quick as he can, 'cause that money for him. No one had no say who he have for wife. But the nigger husbands wasn't the only ones that keeps up having children, 'cause the masters and the drivers takes all the nigger gals they wants. Then the children was brown, and I seed one clear white one, but they slaves just the same." For this, and probably many other reasons, Green remembered Pinchback as "the biggest devil on earth" who "had the St. Vitis [*sic*] dance" and liked "to make he niggers suffer to make up for his squirming and twisting."[39]

Andy Anderson was born in Williamson County in about 1844, property of a man named Haley who owned about thirty adult slaves and twenty children too young to work. Anderson recalled that Haley was "kind to his colored folks, and . . . other white folks called we-uns the petted niggers." Anderson recalled that the plantation was managed "sort of like the small town, 'cause everything we uses am made right there." For example, one slave both tanned hides and made the shoes needed on the plantation. Haley also owned about one thousand sheep, and the slaves sheared, carded, spun, and wove the wool to make their clothes. The plantation's cattle provided "the milk and the butter and beef meat for eating." Haley also raised turkeys, chickens, hogs, and bees. The plantation grew only cotton for sale. Corn and wheat were produced in small quantities for consumption on the plantation. Anderson recalled, "The living for the colored folks am good. The quarters am built from logs like they's all in them days. The floor am the dirt, but we has the benches and what is made on the place. And we has the big fireplace for to cook, and we has plenty to cook in that fireplace, 'cause Massa always 'lows plenty good rations, but he watch close for the wasting of the food." However, when the Civil War began, Haley joined the Confederacy and hired an overseer named Delbridge. At that point, Anderson recalled that "the hell start to pop, cause the first thing Delbridge do is cut the rations. He weighs out the meat, three pound for the week, and he measure a peck of meal. And 'twa'n't enough. He half-starve us niggers, and he want more work, and he start the whippings." Anderson was soon sold to a slave owner named House in Blanco County, described as "another man what hell am too good for." When Anderson accidentally damaged a wagon hauling firewood, House had him tied to a stake, "and every hour for four hours they lays ten lashes on my back. For the first couple hours the pain am awful. I's never forgt it. Then I's stood so much pain I not feel so much, and when they takes me loose, I's just 'bout half dead. I lays in the bunk two days, gitting over that whipping, getting over it in the body but not the heart. No, sir, I has that in the heart till this day."[40]

Ben Simpson gave the Federal Writers' Project an extraordinary account of his life as one of the slaves of a brutal killer named Alex Stielszen. Simpson reported that he was a slave who, with his mother, sister, and several other slaves, was brought to Texas from Georgia by Stielszen. Leaving Georgia in a covered wagon, Stielszen "chains all he slaves round the necks and fastens the chains to the hosses and makes them walk all the way to Texas." Even when it began to snow, Stielszen made his slaves sleep on the ground and refused to let them wrap their feet with rags. Stielszen would beat any slave that fell behind with "a great, long whip platted [plaited] out of rawhide." Near the Texas border, Ben's mother, her feet raw and bleeding, could go no farther. "Then Massa, he just take out he gun and shot her, and whilst she lay dying he kicks her two–three times and say, 'Damn a nigger what can't stand nothing.'" Refusing to bury her, Stielszen pushed on, eventually settling near Austin, where he changed his name and those of his slaves to Simpson. After building a cabin, Simpson became a horse thief and began accumulating his herd in the hills. With almost unbelievable cruelty, he branded his slaves with a hot iron, gave them no clothes to wear, fed them only raw meat and green corn, chained them to trees at night, and refused to bury slaves who died. Alex Simpson soon married a young Hispanic woman named Selena, whom he beat mercilessly when she tried to treat the slaves more humanely. Ben Simpson reported that his owner refused to free him and several other slaves at the end of the war, keeping them for three more years, until he was hung for stealing horses.[41]

The system of plantation agriculture and slavery that developed in antebellum Texas was an outgrowth of and intimately linked with the slave economy throughout the southern United States. Nevertheless, it benefited from the virgin soils and expanding economy of Texas. As a result, Texas slave owners could often produce more crops and livestock per field hand than their counterparts could in the eastern states. Plantation agriculture was expanding and prospering when it was brought to its knees by the Civil War. The principal sources of that prosperity—for both small farms and plantations—were hunting, stock raising, and crop production.

Chapter 8

HUNTING AND STOCK RAISING

As ANGLO-AMERICANS BEGAN TO SETTLE Texas in the 1820s, they found abundant wildlife. The "good old Texas style" of living often involved hunting to provide meat, skins, protection of crops and domestic livestock, and, without doubt, sport. Most travelers to antebellum Texas found an abundance of game, much of which would become scarce as settlers encroached and farming became widespread. Nowhere was wildlife more abundant than near the Texas coast. Describing the earliest settlers' life near Matagorda Bay in the late 1820s, Mary Helm recalled that "at no time were we without large herds of deer in sight in every direction, fish and fowl in the greatest abundance, eggs on the island and in the bay, could be gathered by the bushel, oysters without stint, of the largest size." The deer "had not yet learned to fear man, but would approach him if he would sit down to see what he looked like, and thus come within gunshot."[1]

In 1828 J. C. Clopper crossed the coastal plains between Harrisburg and San Felipe, describing it as "without a tree and scarce a shrub to obstruct the view—it is all clothed with grass from one to two feet in height. . . . This prairie abounds with deer and Mustangs or wild horses—it is beautiful to behold their lofty gambols and wild manoeuvres unconstrained and unshackled by the thraldom of Man." An anonymous traveler reported that near

Anahuac in late March 1831, "wild fowl are abundant hereabouts at this season: particularly geese and [brant], which form vast flocks, whose noise may be heard several miles." He also reported "ducks, pelicans, snipes, cranes, eagles, hawks, buzzards, owls, wild turkies [*sic*], and many smaller birds." Though deer were numerous, sometimes appearing "in large herds," the flat grassy landscape made them difficult to approach. Hunters were "sometimes obliged to approach them through the grass for a mile, stooping almost to the ground and proceeding with the utmost caution." Wildlife was abundant in the woods, where "turkies [*sic*] . . . wolves, bears, panthers, wild cats, wild hogs, foxes, rackoons [*sic*], and squirrels" were found, but wild cats and panthers were said to be "rather scarce."[2]

In the spring of 1836 John C. Duval, one of the few Texans who escaped the massacre at Goliad, wandered eastward across the coastal prairie dodging Mexican patrols and living off the land. During his trip, Duval observed several cougars, wolves, black bears, antelope, wild hogs, rattlesnakes, horned frogs, tarantulas, centipedes, a drove of "six or seven hundred" mustangs, and a roost with "several hundred" turkeys. Near the Navidad River he crossed a large prairie with abundant game. There he frequently had "a thousand deer in sight at a time" and for the first time "saw the pinnated grouse, or prairie hen [probably an Attwater's prairie chicken]." When he first heard the call of the cock, he mistook it for the "distant lowing of wild cattle, some of which were grazing on the prairie." He also saw numerous wild turkeys, "so unused to the sight of man, that they permitted me at times to approach within a few paces of them."[3]

At night, a hunter could "shine the eyes" of his game by holding a torch over his head or by "burning combustibles in a frying pan behind him, resting the handle on his shoulder [with] his rifle ready in his hand." If the hunter held the light properly, the game would stand transfixed by the light, which would reflect from its eyes, betraying its location and giving the hunter time to take careful aim. The principal danger, of course was shooting domestic animals by mistake.[4]

Black bears were "generally found in thick cane brush" and were hunted with dogs. In addition to preying on hogs, they caused extensive damage to corn, especially during and after the roasting-ear stage. Daniel Shipman recalled that during the night "they would come in by numbers, like so many hogs . . . and go out [of the cornfields] before day." As a result, bear hunting was both a sport and a necessity. During one evening and the following morning in 1824, Shipman and other hunters killed and butchered five bears, shot two others that got away, and saw at least eight more. The meat was sewn into the skins for transport back to the homestead on packhorses, which, because of the scent of bear, had to be blindfolded to accept the burden. Bears were also killed with "setting guns." This method involved finding a bear trail and then securing a loaded gun to two posts driven into the ground such that the gun pointed toward the trail "high enough to strike the bear about the center of the body." A string was then tied to the trigger, looped around the end of the stock, and tied to a stake on the other side of the trail. The hunter left "the balance for the bear to do."[5]

In the 1850s Noah Smithwick had a ranch on Brushy Creek in Williamson County. He described the country as "wild and infested with predatory beasts, the most troubling of which were the big gray wolves—lobos—the Mexicans called them." When "wild cattle" were attacked, they would form a ring around the calves, "presenting a line of horns" to fight off the wolves. But the wolves, which "stood three feet high, measuring six or seven feet from tip to tip" could "drag down a grown cow single-handed" if they could separate her from the herd. Unfortunately, Smithwick's tame milk cows associated the farmstead with safety and would run toward it when confronted by wolves. This allowed the wolves to bring down stragglers. After losing several cows and a number of calves, Smithwick "began to treat the lobo family with strychnine . . . until they were pretty well subdued."[6]

DANIEL SHIPMAN'S TALES

In the 1820s Austin's colonists had numerous hogs, most of which ran almost wild through the river bottoms. These, of course, became a favorite food for the bears and cougars (also known as mountain lions, pumas, panthers, and painters). Daniel Shipman recalled that one night after supper on his family's farm on Oyster Creek, "we were all sitting quietly . . . when we heard a hog squeak like it had not got its squeal more than half out before its breath was stopped." His father "gathered a good torch of cane," the boys got their rifles, and "when we reached the lane there stood a large panther with a half grown hog in its mouth." Daniel raised his gun and "let him have the contents, but hit him a little too high." The panther dropped the hog and ran down the bayou, followed by the dogs. After it had run for about

150 yards, the panther stopped on a log, and the dogs kept it there until the hunters arrived. Daniel recalled that with his father holding the torch, "I took a very deliberate aim at him, and, at the report of my rifle, the flash so blinded us that we could not see where he had gone to." After searching up and down the bayou without success, Daniel's father "got a stick and felt down into the water, and there he found him. We took him out and carried him home; he was one among the largest panthers."[7]

Another evening a hog squealed in a canebrake less than half a mile below the field in which Shipman was working. He ran to the house and got his gun and "two puppies not quite half grown." With the puppies close behind him, he ran to a large pond, but the squealing appeared to be in the cane and brush about fifty yards on the other side. The puppies "dashed into the brush," and when Shipman got there, he "heard a terrible rattling and crashing in the brush, coming directly towards" him. As he got ready to shoot, "directly here came both puppies, bulging out of the thicket, as if the bear was just ready to take hold of them." The puppies took shelter behind Shipman just as a bear came out of the brush, stopping within eight feet of him, but the brush was so thick that the bear did not see him. Shipman recalled that "I could see his head and snapped my gun at it." When the gun misfired, the bear "turned and went back to his hog." Shipman "soon heard the pig squeal and . . . could hear him crunching the poor little thing." Within a short time the bear returned to an old sow that he had hurt "so that she could not get away." Shipman "sallied off obliquely to [his] left, and by that means went around the point of the brush, which put [him] in full view of the bear." With the sun going down behind the bear and "the bright sky reflecting beautifully on the bead of [his] gun," Shipman "took a very accurate aim and pulled down on him; he jumped and went off a few steps and tumbled over; he was not very good meat."[8]

Because of their abundance, meat quality, and excellent skins, white-tailed deer were the most valuable game animal. William Bollaert, who visited Texas in 1842–44, noted that Texans had six ways of hunting white-tailed deer: stalking to within eighty or one hundred yards of the animal; fire hunting at night with "an iron pan basket filled with burning pitch-pine with which they shine the eyes of the deer"; waiting "near their trail as they go to water"; making a salt lick near a watering hole, then waiting nearby, often "up a tree"; hunting with "stag or fox or grey hounds—the hunt seldom exceeds two hours"; and keeping "a pet doe with a bell attached to her neck" to entice bucks to come near the homestead. Other game mentioned by Bollaert included turkey, "Mexican hog" [javelina, peccary], fox, opossum, raccoon, skunk, panther [cougar, puma, mountain lion, painter], leopard [jaguar], leopard cat [jaguarundi], several kinds of squirrel, ground hog, wild hog, wild cattle, buffalo, black wolf [lobo and red wolves], prairie wolf [coyote], hare [jackrabbit], rabbit, mink, otter, beaver, and mustang.[9]

Northeastern Texas was also alive with game. Taylor J. Allen recalled that during the 1850s the northern part of the blackland prairie held "native grasses . . . as high as the arm-pits of a man in the valleys, and as high as the waist on the high ground," supporting "game of nearly every kind," including wild horses, deer, buffalo, bears, cougars, racoon, wolves, and coyotes. Turkeys, geese, prairie chickens, quail, and doves were commonly killed for food and were considered pests because they damaged grain crops. As in the Hispanic period, animal skins were valuable commodities. For example, in 1852 Geiger and Company from Shreveport advertised in the *Marshall Republican* that it would pay "the highest market price" for one thousand cougar and five hundred bear skins.[10]

In 1851 Elise Waerenskjold wrote to friends in Norway that "quite a few beasts of prey" were found in northeastern Texas, including cougars, bears, wolves, foxes, opossums, skunks, several types of snakes, and alligators. However, they did not pose a danger to humans, and travelers could safely "sleep in the open either in the wagon or on the ground . . . even though unarmed and far from people, whether it be on the prairie or in the woods." Waerenskjold wrote that there were "still a great many deer about," as well as "numerous smaller animals like rabbits and squirrels (which are very tasty) as well as many wild birds such as turkeys, geese, various types of ducks, prairie chickens, and, in the fall, countless swarms of doves, besides smaller birds." The doves she referred to were almost certainly passenger pigeons, now extinct. She wrote that they "come by the millions; they look like a dark cloud and there is a sound in the sky as if a great storm is approaching. My husband killed about thirty in one shot, and where they roost at night whole wagonloads can be killed." Waerenskjold also noted that chicken snakes, though harmless, could "put a scare into newcomers" by coming into houses in search of eggs. "The reason for its intrusion into houses is that the hens usually have their nests under the beds and up in the lofts."[11]

CATTLE RANCHING

Antebellum Texas' wide expanses of unfenced prairies and forested river bottoms were ideally suited for extensive livestock production. Ranching historian Terry Jordan has concluded that antebellum Texas ranching was derived from two major traditions: the "cowpen" culture of the southeastern United States and the "vaquero" tradition of Hispanic south Texas. In the early 1700s southeastern North America contained large numbers of small, horned, "black" animals, the basis of the southern Anglo "cowpen" tradition of cattle raising. In the early 1700s South Carolinians began to raid the Spanish missions and ranches along the St. Johns and Apalachee rivers in northern Florida, where they stole thousands of the small *criollo* cattle that had long been raised by the Spaniards throughout the New World. The Anglo and Spanish cattle doubtless interbred, and the resulting small "scrub" cattle spread throughout the South. Unlike their contemporaries in Europe and the northern states, southern cattle producers could leave their animals to fend for themselves over large unfenced ranges without fear of large losses to severe weather. "Cow hunts" were conducted at least once a year to earmark and brand the calves. Cattle were regularly rounded up and brought to fenced cowpens. There they were given salt and became accustomed to the herders and cow dogs used to manage them. Houses for the manager and "cowboys" were often located near the pens. Hogs roamed the woods with the cattle. Gardens and crops were grown in fields fenced to prevent damage by livestock. Cowpen owners with thousands of head often lived elsewhere, their herds entrusted to managers. Hands were normally Anglos, but slaves were also common.

After the herds became dependent on the salt that herders spread on grass, tree trunks, and flat stones near the cowpens, they seldom strayed far from the source. In some cases, calves were penned during the day while the cows grazed freely. When the cows returned to their calves at night, they were penned together until morning. Cow dogs were used to herd cattle, and they were indispensable to the cowboys, who, unlike their Hispanic counterparts, used a bullwhip to control and drive the cattle. Fire was used to clear underbrush and stimulate growth of fire-tolerant grasses and forbs. Overland drives were used to move cattle to market and to seasonal pastures. Prior to drives, the cattle were deprived of salt, and it was used to lure them along the trail. During a drive, one or two men led oxen at the head of the herd while others followed on foot or horseback, cracking whips. Cow dogs were used to prevent strays.[12]

Colonial, slave-holding Anglo-Americans in South Carolina and surrounding areas moved westward along two routes: westward through the pine barrens along the Gulf Coast and northwestward to the Ohio River Valley and then southwestward through Missouri and Arkansas. Upon reaching Texas, these settlers established three centers of cattle herding: the southeastern coastal prairies, the pine forests south of Nacogdoches, and the prairies of northeastern Texas.

The westward migration followed the longleaf pine belt that parallels the Gulf of Mexico all the way to western Louisiana and eastern Texas. Throughout the belt, fires were set to suppress undergrowth and stimulate grasses. In addition, streamside canebrakes and isolated prairies provided forage throughout the year. As these herders moved westward, they encountered the French system of *vacheries,* or cattle ranches, that boomed on the southern Louisiana prairies. Spanish control of Louisiana during the last third of the eighteenth century, as well as contacts with Hispanic ranchers in south Texas, strongly influenced the French tradition. By 1770 there were five to seven times as many cattle as people in the prairie parishes, and by the early nineteenth century the ratio had increased to fifteen to one. These French herders expanded westward onto the coastal prairies of southeastern Texas, which in the 1850s were described as "sparsely settled, containing less than one inhabitant to the square mile, one in four being a slave." About 5 percent of the inhabitants were French-speaking "herdsmen, cultivating no other crop than corn, and of that, not enough to supply their own bread demand." Frederick Law Olmsted observed that "they live in isolated cabins, hold little intercourse with one another, and almost none with the outside world."[13]

The coastal prairies of Texas began to attract Anglo-American cattle herders in the early 1820s. Bringing some cattle with them, they must also have purchased cattle from the vacheries of southwestern Louisiana and Mexican ranchers along the San Antonio. They must also have captured orejanos that had strayed onto the vacant lands between the San Antonio and the Brazos. As they moved onto the coastal prairies, these herders began to acquire some of the skills, such as use of the lazo, honed by the Spaniards in northern Mexico and Texas. As a result of importation from the United States, purchase, capture, and natural increase, by 1826 the Austin colony had an estimated thirty-five hundred cattle. That number rose to twenty thousand in 1830, and in 1831 herds of six hundred to seven hundred were reported along the Brazos. By 1837, in spite of the disruptions caused by the Texas Revolution, herds of five hundred to four thousand were reported near Houston, and herds of two thousand or more were common on the lower Brazos. By 1840 there were 50 ranchers in the coastal prairie who owned one hundred cattle or more. By 1850 more than 130 ranchers in the area owned at least five hundred cattle. The great majority of these ranchers had Anglo-American surnames, and many, especially those with large herds, were slaveholders.[14]

A herdsman named Kuykendall may have been typical of small-scale slave owners who settled in the Austin colony. He had come from "beyond the Sabine" with "nothing of this world's goods but a few cattle" in about 1822, settling with his family "upon the borders of the Brazos timber" between San Felipe and the coast in present-day Fort Bend County. Obtaining title to his league of land (one league equals 4,428 acres) in 1824, by 1837, even after having lost large numbers of cattle to Santa Anna's and Sam Houston's armies, he had over two thousand head, worth six dollars each. A traveler noted that "if universal testimony is entitled to credit, stock of the horned kind will double every three years, making an allowance of twenty per cent for annual loss . . . without any more labor than is required to mark and brand the calves. The prairie, both winter and summer, furnishes the most abundant and nutritious pasture, and even salt is not necessary for the stock, as the dew is highly impregnated with the saline properties of the sea." Despite these glowing recommendations, ranching on the coastal prairies could present challenges. Although Texas winters were much less severe than those in the northeastern United States, northers brought cold, wet weather several times each winter, posing a serious problem for cattle living on the prairies without shelter or supplemental feed. Though they instinctively sought "the nearest shelter of trees," on the treeless coastal prairies they fell "by thousands before a freezing rain." These northers could be equally deadly for the traveler, and "teamsters, herdsmen, and travelers, caught out far from habitations, not unfrequently perish[ed]."[15]

In 1831 Taylor White was the wealthiest inhabitant on the lower Trinity near Anahuac. An anonymous traveler reported that White's "estate" was about five miles up the Trinity from Anahuac, and his log house faced north overlooking an extensive prairie, "a little in advance of a tract of woodland, which skirted a small stream." He had lived in the area for only a few years and had between three thousand and four thousand head of cattle. Each year drovers from New Orleans came to the region "to purchase cattle," which they took back "in great numbers." Like other Texas herders, White routinely rounded up his cattle to separate them for sale or to brand the calves. The traveler observed that Texas cattle were "much wilder and more spirited" than those in the United States and were quick to respond to danger. When "occasionally one will begin to bellow," the others "raise their heads, hold their tails almost straight up in the air, and run violently towards the sound," often gathering in the hundreds in response to a threat. The men who managed White's herds had adopted at least some of the vaquero's techniques, using the lazo to work cattle in the pens. They would "throw their lazos over the horns or necks of the young animals, which, in attempting to escape, dash themselves violently upon the ground, and become almost strangled." This experience caused them "to acquire that horror of the lazo that all the tame animals of Texas exhibit." Even wild oxen and horses that had previously been roped and thrown to the ground would conduct themselves "with humility and submission" when they felt the rope on their neck. Refractory milk cows would stand "stock still at the pail with a lazo over their horns, though there was no actual force applied." White, like many other prosperous stock raisers, grew crops and vegetables to complement his livestock operations. Though his soil was poor, "of that sort which is called hog-bed Prairie," it was judged "capable of producing considerable crops even with the least possible labor." Once the land was plowed, corn could be "merely dropped into holes, made with a stick, and [would] grow and yield pretty well even without hoeing." White employed a number of laborers and maintained a "comparatively small and ill furnished" dairy that produced butter, cheese, buttermilk, and fresh milk. The farm also produced vegetables and had "cane and cotton both growing in small patches."[16]

In 1842 Joseph Eve, the United States chargé d'affaires to Texas at Galveston, wrote that when "Taylor White moved here nineteen years ago his whole fortune was three cows and calves, two small [ponies,] a wife and three children." In 1842 he owned "about 40,000 acres of land[,] upwards of 90 negroes, about 30,000 head of cattle," and he had sixty thousand dollars "in specie deposited" in New Orleans. The previous spring he had "marked and branded" 3,700 calves, and in the fall he had sold 1,100 thousand-pound steers, "which he says cost him not more than 75 cents a head" to drive to New Orleans. Eve remarked that "what is extraordinary [is that] he cannot read or write and has made his fortune raising stock alone."[17]

The northeastern Texas prairies began to attract a substantial number of settlers, some of whom were slaveholders, in the 1820s. Though most had been born in Tennessee, Kentucky, North Carolina, and Virginia, they had usually passed through Tennessee, Missouri, or Arkansas before arriving in Texas. The earliest settlers lived along the Red River, but by the early 1830s ranchers began to move onto the prairies on either side of the Sulphur River. These northeastern Texas prairies, found on high ground between forested stream bottoms, were the broad northern extreme of the blackland prairie, a long stretch of heavy clay soils supporting vigorous tallgrass prairies that extended southward in ever-decreasing width to San Antonio. Beginning on the east with a number of small fire-maintained prairies within the forests of the region, grasslands increased in number and extent to the west. Though water for cattle was scarce on the prairies, except in dry periods it could usually be found along the streams. The population of the region grew rapidly after Texas joined the Union, from about seventy-five hundred in 1849 to about twenty thousand in 1850. Cattle numbers on the northeastern prairies also increased, from about 27,000 in 1845 to 54,000 in 1850 and 82,000 in 1855. James Hopkins's cattle numbered only about 50 in 1838, but they had increased to 400 in 1845 and 1,500 in 1854. Meredith Hart owned only 20 cattle in 1840, but he reported 1,000 in 1850. Solomon, Martha, and Daniel Waggoner's herd grew from 30 in 1842 to 440 in 1854.

Northeast Texas herders brought with them the Anglo-American herding tradition, its nomenclature, and its techniques. They ran their cattle on the open range, registered earmarks and brands to identify their cattle, used cow dogs to pen their herds about every two weeks, and provided salt to keep them tame. Their cattle probably varied widely in size and color, and their bloodlines were probably similar to those in the upper South. There were no Hispanic ranchers in the region, and the herders made little or no use of Hispanic ranching techniques or tools prior to the Mexican War in 1846–48. During the 1830s cattle raisers in northeastern Texas trailed their cattle to Oklahoma for sale to the U.S. Army and reservation Indians. In the late 1830s the Red River raft, a large natural logjam that limited boat traffic, was removed, and in the 1840s many herds were driven to both Jefferson and Shreveport, then shipped down the river to New Orleans. Like their counterparts in other parts of Texas, some herders also kept hogs in the river bottoms, raised horses, experimented with sheep production, and produced crops. For example, in 1850 James Hopkins, for whose family Hopkins County is named, owned eight hundred cattle, six hundred hogs, and fifty sheep. He also produced five hundred bushels of corn, six hundred bushels of oats, twenty bushels of rye, and fifty bushels of sweet potatoes. Other cattle raisers also grew cotton and wheat. During the 1840s Hopkins owned from four to eight bondsmen. But slaves were not necessary for cattle production in northeastern Texas. Almost 40 percent of the cattle raisers, some with large herds, owned no slaves; more than 40 percent owned fewer than ten.[18]

The piney woods of eastern Texas were home to a small Hispanic ranching industry throughout the eighteenth century. In addition, between about 1810 and 1830 many of the Indians in the southeastern United States were forced off their lands and moved, often by force, westward to new reservations. In their efforts to assimilate and avoid displacement, many of these Indians had adopted farming and ranching techniques similar to those of the Texians. Eastern Indian tribes, including the Cherokee, Choctaw, Alabama, and Coushatta, moved into the piney woods in the 1810s and 1820s and began to farm and raise cattle. However, beginning in the 1820s large numbers of Anglo-Americans arrived. In 1839–40, during Mirabeau B. Lamar's presidency, most of these Indians were expelled from Texas. In addition, Texians purchased most of the Tejano ranches around Nacogdoches. By the 1850s little remained of the Hispanic or Indian ranching traditions. Texian cattle herders from both the upper and lower South occupied the open pine forests. Their cattle were the breeds they brought with them, mixed to some degree with the criollo cattle raised by Tejano ranchers in eastern Texas and on Louisiana vacheries. Piney woods ranchers adopted the lazo from their Spanish predecessors. As more Anglo-American settlers arrived, the industry grew. In 1840 twenty-seven herders owned more than one hundred cattle. By 1850 almost three hundred herds of one hundred or more could be found in the region, eighty-four stock raisers owned more than two hundred head, and the three largest owners had more than eight hundred cattle each. Most of the eastern Texas cattle herders settled in the sandy longleaf pine forests south of Nacogdoches. The east Texas Indians, by now reduced to very small numbers by European disease and expulsion by Anglo-American settlers, had long used fire to maintain open forests with an understory of fire-tolerant grasses. These lands, so similar to the longleaf pine belt in Georgia, Alabama, and Mississippi, must have seemed like home to herders from these southern states. Naturally, the piney woods cattle raisers found markets in Louisiana and Mississippi. During the 1830s, 1840s, and 1850s many migrants moved along the roads through Nacogdoches to central Texas. In addition, as

English longhorn and Spanish criollo cows

east Texas farms and plantations depleted their soils, their owners often moved west. As a result, many piney woods cattle must have been driven west to establish herds in central Texas.[19]

During the mid-1700s Robert Bakewell developed the first improved English breed. The large, beefy Bakewell longhorn had long, drooping horns that often curved forward beyond the nose. They were long bodied, long legged, and red or brindle in color; usually had a white stripe along the back; and had bad dispositions. Valued for their hardiness, ease of calving, and strong maternal instincts, Bakewell longhorns were brought from England to Virginia in the 1780s. Expanding northwestward, Anglo-American herders took these valuable cattle to the Ohio valley, where they were used to improve the local "scrub" cattle. They were popular in Ohio, Kentucky, and Tennessee until the 1820s and 1830s, when large numbers of families migrated to Texas, often by way of Missouri and Arkansas, driving substantial numbers of cattle with longhorn blood.[20]

Reports from the 1830s suggest that substantial English longhorn blood had been introduced into Texas. A man named Neal moved from Kentucky to a ranch between San Felipe and Harrisburg in 1829. By 1832 he owned between seven hundred and eight hundred head and drove small herds to New Orleans. They probably had English longhorn blood because a visitor described them as "standing very high and of most symmetrical form. The horns of these cattle are of unusual length and in the distance have more the appearance of stag's antlers than bull's horns." In 1837 another visitor observed cattle on the Brazos marked with "rings, streaks, and speckles" and colors as varied "as the hues of the kaleidoscope and as plump and round as the stall-fed ox." In addition, the cows were described as "generally of a fine, large size, equal, if not superior, in this respect to the stock in the United States but admitted to be inferior for all purposes of the dairy." Noah Smithwick recalled "two yokes of

Texas longhorn cow

long-horned Texas steers" pulling the Texans' cannon on the way to San Antonio in 1835. A rancher on the lower Brazos in 1833 owned a steer with horns four feet, five inches from tip to tip. The colors of the cattle in Austin's colony, described as black, black and white, white and black, white, white with red ears, red and white, red, and brindle, also suggested mixed blood.[21]

Despite the mixing of Texian and Tejano herds, the two types of cattle remained recognizable well into the 1840s. Viktor Bracht, a German immigrant and businessman, reported that "the more common one is the American kind; the other, the Spanish or Mexican cattle, is found only in the West. The former is by far the more useful and tamer than the latter, which at times grows wild again." Eventually, however, the interbreeding of criollo, scrub, and improved English cattle, combined with natural selection for aggressive, thrifty, disease- and stress-resistant types, resulted in the famous Texas longhorn, which became a recognizable type before the Civil War.[22]

The Tejano cattle industry along the San Antonio River and south of the Nueces River experienced severe setbacks after 1810, when Mexican patriots such as Miguel Hidalgo y Castilla and José María Morelos began to challenge Spanish rule. Twice, in 1811 and 1813, Texas was taken over by rebels, but they were quickly overthrown by royalists, who took revenge on those, including ranchers, who had supported the rebels. With each turn in the fortunes of revolution, ranches bore the brunt of retribution. Filibusters and common criminals from the United States added to the confusion and provoked additional retribution from the Spanish government. From 1818 to 1821 military protection for Texas was almost entirely absent, and the Comanches, recognizing the settlements' vulnerability, harassed outlying ranches, stealing cattle and killing those who could not defend themselves or flee. In the 1820s the new Mexican government, facing challenges closer to home, neglected the defense of the northern frontier. The result was that many ranches in both southern and eastern Texas were abandoned, and the ranch families retreated south across the Rio Grande, to San Antonio, or eastward into Louisiana. In the early 1830s Jean Louis Berlandier reported "few dwellings between Río Grande and Río Nueces" that had not been "watered in the blood of some traveller or colonist." Before the last years of Spanish rule, "numerous huts, surrounded by fields and herds, dotted the now deserted lands," and where there had been "several hundred thousand animals of all kinds in the jurisdiction of Camargo," mostly north of the Rio Grande, "all the herds together scarcely number[ed] twenty-five thousand head." Around San Antonio, there were few remaining cattle, and because of Indian raids it was too dangerous to cultivate the fields near the town.[23]

The Tejano ranching industry was further disrupted by the Texas Revolution. Herds strayed or were driven off as General Santa Anna's army pursued the Texians toward the war's final battle at San Jacinto. The confusion was compounded when the Texians moved back westward after the battle. Eager to establish their own ranches, and suspicious that most Tejanos were Mexican sympathizers, Texian ranchers and cowboys rounded up Tejano cattle in the confusing aftermath of the war. After the revolution Texian ranchers and planters put pressure on Tejano ranchers, especially in the piney woods around Nacogdoches and in the region between Goliad and San Antonio. Tejano ranchers, unprepared to resist the economic, political, legal, and sometimes violent pressures of land-hungry Texians, began to lose control of their herds and ranches.

Texians, already prejudiced against Tejanos because of the revolution, viewed the remaining Hispanic ranchers as, at best, second-class citizens and more commonly as unwelcome Mexicans. In some cases the remaining Tejano rancheros were eager to sell their livestock and land. In other cases the cowboys simply took what they could. Soon, groups of Texian cowboys were driving herds of cattle, horses, and mules eastward to start herds of their own, sell to new immigrants, or drive them all the way to the markets of New Orleans. The bankrupt Texas government, a rapidly increasing population, land speculation, and lack of banks and hard currency made trade in stolen livestock easy and lucrative. Conflicts increased after Texas joined the Union in 1845, and, in opposition to Mexico, the United States pursued Texas' claim to the lands between the Nueces River and the Rio Grande. The number of Anglos in south Texas increased during the Mexican War when the U.S. Army and the numerous civilians needed to supply it passed through south Texas and set up camp on the Rio Grande. The U.S. Army used local herds for food and mounts. Again, many Tejano ranch families fled south during the war, often leaving many of their cattle behind. This made it easier for Texian cowboys to round up the remaining herds and drive them north.[24]

Between 1845, when it joined the Union, and 1860 Texas experienced rapid growth. By 1860 the frontier of settlement had moved westward to a line extending from Maverick County on the Rio Grande north-northeast through Edwards, Menard, McCulloch, Coleman, Eastland, Young, and Archer counties, to Wichita County on the Red River. During the 1850s the cattle industry was still expanding in east Texas (east of the blackland prairie and north of the coastal plains) and in the coastal plains (northwest of Victoria). However, it was growing even faster to the west on the prairies of north-central Texas (the cross timbers, grand prairie, and blackland prairie north of Austin) and south-central Texas (the prairies from Austin east to Brenham and south to Kingsville). Many Texian ranchers from the southeastern coastal prairies migrated into south-central Texas, where there were fifteen cattle for every person in 1860. Ranchers from the prairies of northeastern Texas and the pine forests south of Nacogdoches moved into the prairies of north-central Texas. For example, in 1850 the southeastern coastal prairies and east Texas had over 575,000 cattle, and there were about 350,000 head in the north-central and south-central regions. By 1860 the cattle population of east Texas and the coastal prairies had increased to over 850,000, but herds in the north-central and south-central regions had exploded, to approximately 1.9 million head.[25]

PRAIRIE FIRES

The perennial grasses that dominated the Texas prairies provided good grazing during the warmer parts of the year, but in the late summer and autumn their stems elongated, and they produced seed. These stems were low in protein and high in fiber, reducing their value as forage. To clear the prairie of this low-quality material, both the Indians and Texians burned the prairies. In 1846 Ferdinand Roemer observed "the beautiful spectacle of a prairie fire" at night near Torrey's trading post on the Brazos near present-day Waco. He described it as "like a sparkling diamond, the strip of flames, a mile long, raced along over hill and dale, now moving slowly, now faster, now flickering brightly, now growing dim." The travelers assumed that the Indians had started the fire, "since they do this often to drive the game in a certain direction, and also to expedite the growth of grass by burning off the dry grass." Olmsted reported that he "passed a man engaged in firing the prairie. He drew a handful of long, burning grass along the dry grass tops, at a run. Before the high gale it kindled furiously, and in fifteen minutes had progressed a mile to leeward, jumping, with a flash, many feet at a time. In a moderate wind we had once noted the progress of prairie flame, to windward, at about one foot per minute."

Although prairie fires stimulated grass growth and suppressed woody plants, they could kill wildlife. In the spring of 1836 John C. Duval, crossing the coastal plains near Caney Creek, found himself in the path of a rapidly advancing prairie fire. Using his flint and steel to ignite some tinder, he lit the dry grass and set a backfire. Safe in the burned area produced by his fire, he watched "the bright tongues of flame flashing out at intervals through the dense column of smoke." As he watched, "hundreds of deer, antelope and other animals came scampering by [him] in the wildest terror, and numerous vultures and hawks were seen hovering over the smoke, and occasionally pouncing down upon rabbits and other small animals, roused from their lair by the advancing flames."[26]

In the mid-1850s California, in the midst of the gold rush, was an attractive market for Texas cattle. In 1854 Olmsted observed "a California cattle-train" composed of "four hundred head of oxen, generally in fine, moderately fat condition" with twenty-five men to drive them. Each man was mounted on a mule and was well armed with a "short government rifle and Colt's repeaters." Their provisions were carried in two large wagons and a cart, and they were accompanied by "a French family, which was very comfortably fitted for up to six months' residence and conveyance of a woman and several children." Olmsted observed that the economics of driving cattle to California were favorable, with oxen costing about fourteen dollars a head in Texas and selling for one hundred dollars a head in California. In addition, if market prices in California were temporarily low, the cattle could be pastured there, gaining weight and value until prices recovered. Olmsted was told that about four men were taken for each hundred cattle, but only the older, more experienced drovers drew a wage. The younger men were migrating to California and were happy to spend five or six months driving the herd in exchange for having their expenses paid traveling in the company of a well-armed group.[27]

HOGS

Hogs were another important component of the southern Anglo tradition. Like cattle, they were allowed to range freely. Taylor J. Allen, an early resident of Fannin County in northeastern Texas, described how his father kept "several hundred, long snouted, long tusked, back windsplitter hogs." They "kept fat in those days, as there was always plenty of persimmons, hawes, grapes, pecans, hickory-nuts, walnuts, etc." In order to distinguish its hogs from those of others, the family earmarked (cut a distinctive notch from the ears of) piglets. They "would catch the young pigs to mark them where the old mamma sow had left them in their snug little beds in the high cane brakes on leaves and grass." When the piglets were marked, they would squeal, bringing a "rush of the vicious herd of swine as they would come in defense of their litter." The hogs "killed and mutilated" many of the family's dogs, and family members were "often compelled to climb a tree or use some decoy or strategy to induce them away so they would do us no injury." When the family needed pork, the father would catch a hog "when he was apart from the herd," cut a slit in the hog's snout, tie a rope through it to "our old horse, Selim's tail," and "the horse would drag the hog by his tail without any apparent inconvenience or injury." Though most Texas farmers probably used methods similar to those described by Allen, others were more careful. In 1851 Elise Waerenskjold recommended that sows be kept penned for several weeks after farrowing because "if one lets them run about in meadows and woods with their young while these are quite small, the increase will be very slow because most of the little ones will be killed."[28]

Olmsted wrote that hogs were plentiful in the bottomlands of east

Wild hog

Texas, and though they were claimed by planters, poor white farmers did not hesitate to kill them for their own use. "If there ever were any hog-thieves anywhere," said a planter named Strather who lived on the Sabine River, "it's here." In years with little mast production the hogs went hungry. For example, Olmsted reported that during the winter of 1853–54 the hogs in the Trinity bottoms became "perfectly frantic and delirious with hunger," rushing in to eat corn from under the noses of his horses and running directly through a campfire to steal a chicken on a spit. Only by "procuring an excellent dog" and pitching their tents "*within* a large hog-yard, putting up the bars to exclude them," could Olmsted's party protect themselves from this "disgusting annoyance."[29]

On C. W. Tait's plantation, hogs to be slaughtered in the winter were captured and penned in September, then "fed on corn previously shelled, and soaked in water for two or three days." Pumpkins and sugarcane were also fed to the hogs. Affleck's *Southern Rural Almanac* recommended that plantations begin to raise improved breeds of hogs to improve their supply of meat. Although recognizing that wild hogs could take better care of themselves than improved breeds, it asserted that "a few good animals, well cared for, will yield much better results than large stocks running at large." The almanac highlighted breeds such as Woburn, white Berkshire, black Berkshire, and Suffolk, and it recommended that plantations use "an old or weakly hand, to attend to a good but moderate stock of [improved] hogs. . . . If such a hand put, each season, twenty-five head of hogs into the meat-house, weighing two hundred pounds nett [*sic*], he would clear more than he possibly could in the cotton field." Breeding stock could be purchased from Robert Scott's Locust Hill Farm in Frankfort, Kentucky.[30]

MUSTANGS

Herds of wild horses, called *mesteños* by Hispanics and *mustangs* by Texians, provided a source of both mounts and income. F. S. Wade gave the following account of a mustang hunt organized by his uncle Add in 1828. The hunting party consisted "of eight men all splendidly mounted and my nigger Jim for cook." They used a pack mule to carry "about forty lariats and hackamores [halters], some axes, etc." They went to the "[San] Gabriel country, now Williamson, Milam, and Burleson counties, this being among the best mustang ranges in the province of Texas." Pitching camp "on Mustang [Creek] about two miles below the present city of Taylor," they constructed a pen out of "elm poles built eight feet high, enclosing about half an acre, a gap on one side with brush wings widening out from the gap." Wade recalled "no mortal ever saw a prettier country. It was in May and the grass was as green as a wheat field. The south wind made it wave like the sea. There were patches of buffalo clover that were as blue as the sky, then spots of red and white posies that filled the air with sweet smells. Herds of mustangs off to the east, buffalo, antelope, deer everywhere." They picked out a herd of mustangs with more than one hundred head "led by a big sorrel flax-mane- and-tail stallion." Wade and another man watched the herd for several days to determine its range, and after the pen was complete, Wade stationed men at strategic locations. After a delay of several days caused by Indians in the area, the mustang hunters started to "walk down" the herd, taking turns following the horses to exhaust them. "We walked down them mustangs in twenty-four hours, got them all in the pen; and they was sure a fine lot." In the herd they found "ten head of broke horses and mules that had got away from settlements, and a big brown horse with a Spanish brand" that Wade afterward heard had gotten away from a Mexican general. They roped and separated about sixty of the best horses and turned the rest loose. Wade recalled "it took about two or three days to gentle our stock; then we tailed them six in a string and returned to the settlement" where they "had no trouble selling out at from $30 to $60 a head."[31]

SHEEP

Although not nearly as important as cattle or hogs, both Mexican and European breeds of sheep were raised in Texas. The *Texas Almanac for 1859* reported that the first attempt to improve "coarse wooled Mexican

Merino ewe

sheep" was made by Thomas McKinney, who crossed them with imported Bakewell, Merino, and South Downs breeds on his ranch in Fort Bend County in the early 1840s. However, "for many years he was unsuccessful." He found that the crosses did not do well on his "hog-wallow prairie lands." The "rank coarse sedge grass . . . did not suit sheep." As a result, in about 1850 he moved to a "beautiful place on Onion Creek, near Austin" where he had "uninterrupted success with his sheep." In 1849 the *Texas State Gazette* reported that over 1,200 full-blooded Merino sheep had "just been driven in from Illinois." The wool from these sheep, the property of Clinton Harris, had sold for thirty-seven and a half cents per pound the previous season. Destined for pasture in Grimes County, the flock was expected to prove "a great desideratum to Texian wool-growers." In 1850 a number of cattle herders in northeastern Texas also kept a few sheep. For example, the Waggoner family owned 350, James Hopkins kept 50, and Jason Wilson had 30. Because of their lack of contact with southwestern Texas, these sheep were probably English breeds from the Ohio valley or Missouri. In 1851 Elise Waerenskjold wrote that sheep raising in northeastern Texas was "very profitable," but the flocks had to be "locked in every night to keep them from being killed by wolves."[32]

In 1852 George Kendall imported 18 Merino rams from Vermont and began using them to improve a herd of 600 Mexican sheep on a ranch a few miles up the Nueces from San Patricio. Though the flock was "healthy enough," Kendall did not like the country, and in early 1853 he moved the flock to "the hilly regions of Comal County, above New Braunfels." He felt that this area, "high, dry, coated with short grass, and without any low or hog-wallow prairie," would be "admirable adapted to sheep raising." The first several years in Comal County were difficult. Kendall lost sheep to a severe snowstorm in the spring of 1853, to a prairie fire in the winter of 1855, to a cold, wet period during lambing in the spring of 1855, and to an extended period of cold, wet weather during the winter of 1855–56. Despite these difficulties, his herd multiplied, and by continually crossing young ewes with Merino rams, Kendall improved the quality of the wool. Although the original Mexican ewes produced only about a pound per year of coarse wool worth twenty cents per pound, full-blooded Merino rams produced eight pounds of fine wool worth eighty cents per pound. Kendall's third generation of crossbred wethers, with seven-eighths Merino blood, produced seven pounds of fine wool per year. After they were shorn in the fall, fat wethers could be sold in San Antonio for mutton, at $2.50 to $3.00 each. As a result of his breeding program, between 1856 and 1858 Kendall's flock doubled in size, tripled in the yield of wool, and quadrupled in the value of wool it produced. He thought that several factors made sheep raising in western Texas a very profitable business. The cost of land was only one to two dollars per acre, compared with twenty to sixty dollars in the northern and western parts of the United States. In addition, Kendall felt that "the two great scourges of flocks, almost the world over," foot rot and scab, as well as "that worst of all epidemics among sheep—the liver rot," would never cause much loss to "flocks in this high and dry region." Another advantage was that, unlike sheep in the northern states, Texas flocks with half or more Merino blood did not need to be confined and fed during the winter. Kendall thought that any sheep with "a coat of five or ten lbs. of wool upon his back . . . will keep healthier in an open enclosure, during the winter, than under a close covered shed." Pens should be built on the sloping south side of a hill and have a stone wall on the north, west, and east sides to break the cold wind. A supply of hay for use in cold, wet weather was also recommended.[33]

Thomas Decrow was a successful sheep raiser whose ranch was at the tip of the peninsula forming Matagorda Bay. He thought that for successful sheep production in Texas, the "only requisite [was] a dry soil, free from fresh marsh and flat and muddy land, with good water and short grass and weeds." In contrast to Kendall, Decrow was more concerned with meat than with wool. He recommended crossing Mexican ewes with South Downs rams, which he found to be good breeders that produced large offspring with wool of medium quality. Beginning in 1842 with 13 Mexican ewes, he gradually built his flock to 1,200 in the spring of 1854. In September of that year he lost about 700, but by 1858 he sheared 1,032, averaging over four pounds apiece. Unlike Kendall, he never separated his rams from the ewes, considering it "better to let the lambs come about the middle of February or later" after the grass had begun to green up. Decrow also took pains to wash the sheep's fleece before shearing. His shearers took his sheep to a freshwater pond where they soaked and washed the fleece, then finished them with clean water from a hogshead. After being left in "a clean portion of the prairie for at least six days," the sheep were sheared.[34]

In the spring of 1854 Olmsted encountered a Mexican shepherd driving a flock of about 500 sheep belonging to a rancher named Caldwell "to one of the islands of the coast." Part of a larger group of about 7,000, the flock had been purchased near the Rio Grande and taken "first to Corpus Christi, afterwards to the upper part of Goliad County." They were described as "Mexican, very poor and thin, coarse-wooled, large-framed, long-legged, without wool upon their bellies, legs, or heads." A few goats were included in the flock, according to Olmsted, as was "the Mexican custom, for the prevention of disease." For lack of protection from the cold, wet northers, 1,500 of the flock of 7,000 had died of exposure. Olmsted also saw 300 sheep belonging to a German farmer in Medina County. They were also Mexican sheep, but in contrast to Caldwell's flock, they were "in process of rapid improvement by admixture of Saxony blood." Provided with shelter at night and with hay during cold, wet weather, "not one had been lost by exposure."[35]

MULES AND OXEN

Prior to the Texas Revolution, mule trains transported almost all merchandise between Texas and Mexico. A traveler described a train of fifty or sixty mules accompanied by a Texian owner and several well-armed Mexican attendants. Each mule was "loaded with different articles to such a degree" that the traveler "was astonished at their being able to travel." Each carried a *carga,* a load of over three hundred pounds on a packsaddle, "bound on so tight as almost to stop their breath." The mules walked "without bridle or halter . . . with perfect regularity in a single line, except now and then a couple of vicious ones were tied side by side." The business was "attended with a variety of hazards and dangers," but goods from the United States were allowed to enter Texas

Mule

without duty to stimulate growth of the territory, and if internal Mexican duties and Indians were avoided, excellent profits could be realized. As a result, mules used to transport goods were "very valuable," selling for fifty to one hundred dollars apiece. With the expense of raising them "very trifling," breeding and selling mules was a profitable business, "considered one of the most lucrative, under proper management, which could be devised in Texas." One Texian breeder named Barrow raised about seventy mule colts every season.[36]

Oxen were used for a wide variety of work, including plowing, skidding heavy logs, turning the screws of cotton presses, and pulling wagons loaded with bales of cotton to market. One advantage of using oxen was that they required minimal care. Taylor Allen recalled that in the evening the oxen were "turned loose with a bell on the leader to roam at will over night, and in the morning would be rounded up, harnessed, and put to work till 11. Then about 1 or 2 o'clock would resume the work of the day preparing the ground or harvesting the crop." However, oxen were notoriously difficult to manage. For example, in the 1930s former slave Bill Homer asked an interviewer, "You never drive the ox, did you?

The mule ain't stubborn 'side of the ox. The ox am stubborn and then some more. One time I's hauling fence rails, and the oxen starts to turn gee [right] when I wants them to go ahead. I calls for haw [left], but they pays this nigger no mind and keeps agwine gee. Then they starts to run, and the overseer hollers and asks me, 'Where you gwine?' I hollers back, 'I's not gwine, I's being took.'"[37]

The unique Texas landscape, with its virgin soils, extensive unfenced prairies, and abundant wildlife, rapidly expanding markets, and herds of near-feral cattle, horses, and hogs, made it a paradise for stock raisers. The abundant, nutritious prairie grasses and mild climate allowed herds to increase rapidly. By the 1840s Texian cattle raisers, like their Hispanic predecessors, were driving herds to Louisiana. As more and more families moved to Texas, the number of farms increased, and the market for breeding stock remained strong. As Texians pushed the frontier into the north-central and south-central parts of the state, cattle numbers increased, and by the 1850s Texian herds were being driven to California. British breeds were imported from the United States and crossed with hardy criollo stock, eventually producing the famous Texas longhorn. Horses, oxen, mules, and hogs were also important to the Texas economy, and a small number of sheep raisers began to use European breeds to improve the coarse-wooled Mexican flocks. Livestock and wildlife were extremely important to the Texas economy, but most Texians produced both livestock and crops.

Chapter 9

COTTON, CORN, SUGARCANE, AND WHEAT

COTTON, CORN, SUGARCANE, AND WHEAT were the most important crops in antebellum Texas. Most of the corn was consumed locally, whereas most of the cotton and sugarcane was exported, usually through Galveston and other coastal ports. Cultivation of cotton and corn was concentrated along the rivers, especially in the valleys of the Trinity, Brazos, Colorado, and Guadalupe. Sugarcane, because of its long growing season and great nutrient requirements, was confined to "a few strips of the richest alluvial soil in the lower coastal country, particularly in the bottom of the Brazos and of a few smaller rivers, such as the Caney, the St. Bernard, etc." Wheat, though grown in small quantities by German immigrants in the San Antonio region, became an important crop when settlers from the northern and midwestern United States began to grow it in the northeastern counties.[1]

COTTON

In antebellum Texas, as throughout the South, cotton was king. It was the major source of income, especially for slave owners, and its planting, cultivation, harvest, and shipping determined the rhythm of work on many farms and plantations. Thomas Affleck recommended that the first labor of the new year should be to clean and repair the "gin-house," a two- or three-story wooden building in which the cotton lint was separated from the seed. In January the gin stand, which separated the lint from the seed, and press, which compacted the lint into bales of four hundred to five hundred pounds, should be oiled and covered with cloth to keep out dust. Preparation for planting began in February by selecting the seed. Affleck advised the farmer to "pick over sufficient cotton seed to plant a part of the crop, from which to select the seed of the year following; without some such practice, all seeds deteriorate." He recommended planting a third to half the cotton crop between the tenth and twentieth of March and the remainder before the end of April. But he refused to recommend specific planting methods because "hardly any two planters have precisely the same manner of planting either corn or cotton . . . although, in all farm work, that which is worth doing at all, is worth doing well; there is none other to which this applies so strictly as in putting the seed in the ground." I. T. Tinsley recommended that on the level coastal plain soils of Brazoria County, cotton should be planted during the last half of February on sandy soils and a month later on heavy clay soils. The seed should be planted in a furrow on the top of "a well-thrown-up ridge, . . . covered over with a light harrow, after which a good

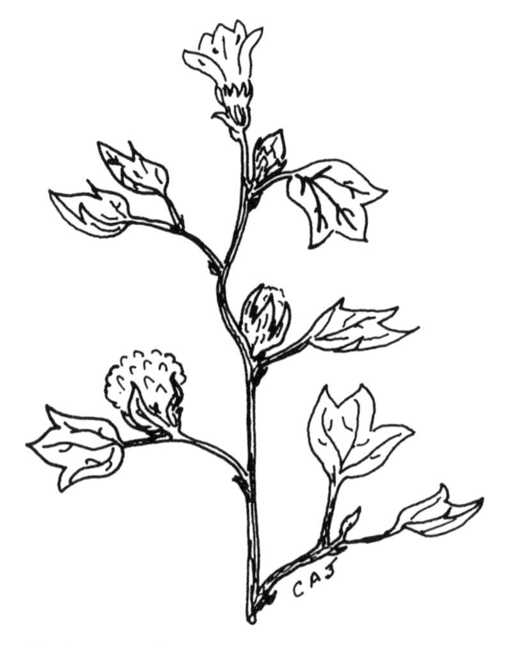

Branch of cotton plant

heavy roller should be run over to press the ground well to the seed." On Anson Jones's Barrington plantation cotton was planted between late February and mid-April, depending on the weather. In contrast, a prosperous planter near Austin favored planting later, "in May, and even in June, as there was no danger of its not maturing here as in the Eastern and Northern cotton states, and its growth is more rapid if not exposed when young to checks from cold."[2]

Because cotton grew slowly, farmers worked hard to keep weeds in check, plowing between the rows and hoeing, usually referred to as "chopping" or "scraping," weeds that grew near the rows. In addition, to guard against poor stands, farmers sowed more seeds than were needed to produce a crop. As a result, hoes were used to thin the cotton plants. Anson Jones began to plow, scrape, and chop his cotton as soon as the seedlings (and the weeds) had emerged, and he usually weeded it two or three times before "laying it by" in June or July. Affleck cautioned farmers that in May, "permit nothing to interfere with the early scraping of the cotton that may yet need it." In addition, plows were used to "mould" or throw soil around the bases of the young plants "as closely as possible behind the hoes; the injury that the young plant receives, by the partial exposure of its roots to the sun's rays, far more than counterbalances any supposed gain in the tending, by the postponement of the moulding." In June, cotton required constant attention to control weeds. In addition, "preparation should be made for cotton picking; sacks and baskets made, if not already done; scaffolds arranged; and every other arrangement made that may facilitate the great business of the fall." In July, "plows, sweeps, or cultivators [should] . . . be kept constantly going" to keep the cotton weed-free "to encourage late growth and the retention of forms [buds] and bolls which are otherwise apt to drop off." In addition, "every requisite provision for cotton-picking should be made by the first of the month, that no time may be lost and nothing hurried when the crop opens—sacks and baskets made and marked, beam and scales properly arranged, gin-yard cleaned off and scaffolds erected, etc." In August the cotton bolls began to open, and picking should begin "so soon as the light hands can gather from 40 to 50 pounds each per day." Arrangements should also be made for hauling the seed cotton to the gin house in order to avoid having the hands carry their loaded baskets on their heads.[3]

September was a critical month for harvesting the crop. Affleck exhorted his readers that "not a day nor an hour of favorable weather should now be lost; but push the gathering in of the cotton crop with all hands—no time now for extra jobs." Despite the need for haste, Affleck cautioned that the quality of the final product was paramount. He recommended that cotton not be picked wet because "it will dry more quickly on the stalk than any where else." In addition, the trash should be removed; "otherwise a prime article cannot be sent to market." The gin stand and cotton press should be maintained and used correctly because if the lint was "put carelessly in the press, rolled up into small tight wads, and trodden down by dirty feet [or] the bale lop-sided and badly covered and tied," the price would be reduced.[4]

The intense effort to pick, gin, and bale the cotton crop continued through October. Affleck insisted that "not a day should be lost" to avoid "violent changes of weather." He was adamant that cotton picking should be completed in November, stating that "if the cotton growers put in no more crop than could be gathered up to the last day of November, they would hold the control of the markets in their own hands, and would have leisure for other work, now neglected or slighted."[5]

Because of the shortage of labor to harvest the crop, cotton yields were usually calculated in "bales to the hand," meaning the number of bales a field hand harvested during the season. Frederick Law Olmsted reported yields ranging from four and a half (in a dry year) to ten bales per hand. Of course, it was often easier to plant than to harvest the cotton, and fields were sometimes left unpicked. In January, near Centerville in present-day Leon County, Olmsted reported that his party "passed one field white with excellent cotton, entirely unpicked." The owner informed them that this was not unusual. "The crop was so great that the hands that had sown the seed were unable to reap [it]." He observed a similar situation a year later in the bottoms of the San Marcos River, where not more than half the cotton had been picked in mid-January. Of course, yields varied in response to weather, soil fertility, the owner's dedication to the crop, and, apparently, his veracity. Olmsted reported that a plantation not far from the Sabine "made eight bales to the hand." A neighbor "who knew the place" revealed that "they had planted earlier than their neighbors, and worked night and day, and he believed had lied, besides."[6]

The first step in getting the cotton crop to market was to separate the cotton fibers, or lint, from the seeds, known as ginning the cotton. Daniel Shipman recalled that in the 1820s his family would pick cotton during the day, "and at night we would put a pile of it on the hearth near the fire, and make it quite warm; had a little machine called a hand gin, made by shaping a small piece of wood—a block about ten inches by twelve, with legs, bench like, and is similar to a washing machine wringer of modern invention. One would feed the cotton into its sylinders [*sic*], while the other would receive it free of seed and place it in a basket near by, and our mothers knew how to weave double weaving, as Jeans, counterpanes, coverlets, etc."[7]

But hand gins had a very limited capacity. This bottleneck in cotton production began to disappear when, in about 1825, the first animal-powered gins based on Eli Whitney's design arrived in Texas. Soon these modern gins were operating near San Augustine, in Austin's colony, and near the coast at Matagorda. Initially brought from the southern states, they were soon constructed locally by blacksmiths. By 1828 there were four or five gins in Austin's colony, and by 1833 the number had probably increased to fifteen to twenty. After the Texas Revolution, the number of gins increased rapidly. Because a gin could normally produce about two hundred bales a season, we can estimate that Texas had about 240 gins by 1850 and more than 2,000 gins by 1860.[8]

The machine used to separate the lint from the seed was called the gin stand. It was housed in a gin house, usually a two-story frame building with the gin stand on the second floor. Power to run the gin stand was usually generated on the first floor by a pair of horses, mules, or oxen harnessed to two poles attached to a horizontal drive wheel about ten feet in diameter. The drive wheel transferred power through a bevel gear to a vertical wheel about seven and a half feet in diameter. A belt on the outer edge of the vertical wheel passed through the floor of the second story and around a pulley on the gin stand. The gin stand was composed of a sturdy frame holding a horizontal shaft with the pulley on one end. About fifty circular saws were fitted approximately one inch apart to the shaft. These saws revolved between openings in a slotted guard called the breastwork or ribs. As the cotton was fed onto the breastwork, the spinning saws pulled the lint through its slots. But the slots were too narrow for the seeds to pass. As the lint was torn away, the seeds fell away from the breastwork. As the lint passed through the breastwork, a cylinder of stiff bristle brushes turning in the opposite direction lifted the lint from the saws and blew it away from the stand, usually into the lint room located on the first floor and separated by a wall from the drive wheel and livestock. Variations on this plan included three-story gin houses with the third story serving as a storage room for the seed cotton, which could be fed down a chute to the gin stand. Some gins used waterwheels, animal-powered tread wheels, or steam engines to generate power.[9]

In the 1820s lint was packed into bags up to nine feet long for shipment. However, by the early 1830s the screw press came into widespread use because it enabled the grower to produce uniform bales weighing from four hundred to five hundred pounds. The press consisted of a

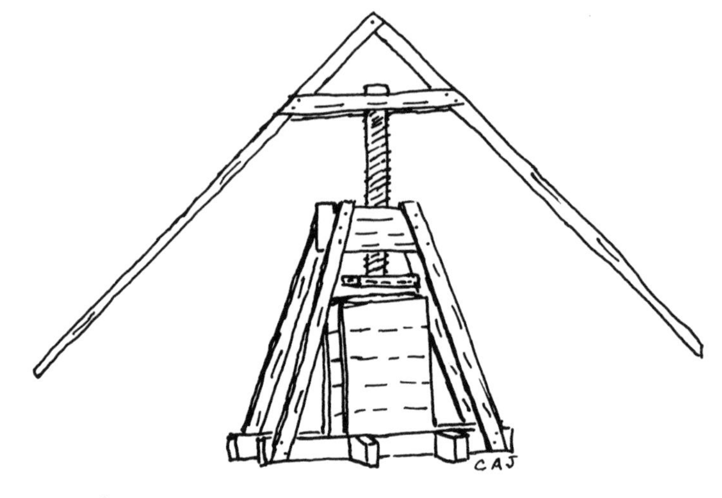

Cotton press

sturdy bale box and a hand-chiseled vertical screw, usually of oak, that forced a block downward into the box. On the upper end of the screw were attached two long beams, called "buzzard wings." These angled downward and were hitched to draft animals that walked in a circle, raising or lowering the screw. To make a bale, the crew raised the screw and pressing block, positioned several ropes and the cloth bagging that would secure the bale in the bale box, tramped the lint into the box until it was full, and placed the top bagging over the lint. The screw and pressing block were lowered onto the lint, and the animals turned the screw until the cotton reached the desired density. The ropes were then tied, and the bale was removed from the press. Gins of this type usually required about ten hands to continuously unload cotton from wagons and carry it to the storage room, drive the draft animals providing power, operate the gin stand, and press the bales. Small plantations used a crew of four or five, stopping the gin stand while the lint was being baled. By the late 1830s Texas planters could buy gin stands from a number of competing manufacturers. By the late 1840s gin stands consisting of forty to sixty saws could be purchased for three to four dollars per saw. The *Texas Almanac for 1857* advertised the "Star Cotton Press, invented and manufactured by M. L. Parry, Galveston, Texas" for forty dollars. The press was advertised to "exceed all others" because it was driven by the same belt used to power the gin stand, it was designed to work under the same roof as the gin, and "Negro women and children [could] operate it without the least danger of getting hurt." It would "last for generations, without repairs," and it could "pack in the very best style 40 bags per day."

In 1856 the cost of a gin house, gin stand, press, mules and harness, and other tools was estimated to be eight hundred dollars. By 1860 the *Texas Almanac* carried advertisements for cotton sweeps, cotton scrapers, cotton harrows, single and double cotton gins, and cotton presses, sawmills, portable and stationary steam engines, gristmills and flour mills, grass and grain scythes, saddles, harnesses, pumps, rifles, pistols, shotguns, and other equipment needed on the farm or plantation.[10]

Affleck's almanac for 1860 also advertised a wide variety of products to help the planter gin the cotton crop. If the cotton press needed a new screw, the planter could purchase one of "Newell's Cotton Screws, of 6, 7, 9, and 11 inches in Diameter, by 12 feet long, and geared for either Horse or Steam Power" from Samuel H. Gilman, New Orleans. The same source also carried "shafting, couplings, pulleys, [gin] stands, hangers, and boxes of all sizes" to equip or repair the gin house. The machinery could be oiled with "Superior Machinery Oil" made from pine resin and available from the Southern Oil Company of New Orleans. This oil was described as "the first ever made from Rosin that was suitable for machinery," and it was offered to the public "at 25 per cent below the Cheapest Lubricating Agent in use." Testimonials included in Affleck's almanac revealed that the "rosin oil" was equal to or better than "the best lard oil" and was being used by railroads and sugar mills.[11]

CORN

Corn was second only to cotton in importance to Texian farmers. The grain was used for food and animal feed, and the fodder was prized for its production and quality. Olmsted reported that the first year after breaking the Texas prairies, weeds were a minor problem, and "one hand is said to be able to tend twenty acres of corn and ten of cotton." In contrast, on Virginia plantations ten acres of corn per hand were "considered a good allowance." Corn planting in south-central Texas began in mid-February. Tinsley suggested planting corn in the "water- furrow," between the beds formed during soil preparation. This allowed plows to clean weeds from between the rows and throw soil around the corn rows without forming "too high a ridge" by the time the crop was "laid by." Corn "so planted" would "always stand a drought better," but in soils that tended to be cold and wet in the early spring, corn was planted in shallow furrows on top of the ridges to prevent drowning the sensitive young plants.

Affleck recommended that corn be planted for both grain and fodder. For grain, he suggested planting it in hills, but for fodder it should be planted in rows two and one-half to three feet apart. The seed corn was first tarred and treated with ashes to prevent birds, insects, and other animals from eating the germinating seeds and seedlings. The tar was applied by soaking the seeds, adding tar mixed in hot water, stirring to thinly coat the seeds, draining, rolling the seeds in ashes, and then drying them before planting. Affleck assured his readers that "neither birds, squirrels, racoons [*sic*], nor even cut-worms, will injure corn thus treated."[12]

Affleck recommended that corn for fodder be planted at two-week intervals from February through May. "No one can form an idea of how great is the advantage to the stock of a place, the having plenty of such green fodder, as a few acres of drilled corn will yield, until they try it. The wonder then is, how they have previously done without it." Corn for grain was planted in hills of about three seeds each. As soon as the young plants had emerged and were growing well, the crop was weeded with a cultivator, plow, or hoe—"lose not a day with it, when the ground is in proper condition." Corn was normally weeded with plows and hoes about three times between planting and the time the crop canopy shaded the soil between the rows, usually in late May or June. The crop was then considered to be "laid by," and no more weeding was required. In July corn fodder was harvested by stripping, or removing, the top part of the plant above the ear. If the corn was not stripped too early, the leaves below the ear would provide enough photosynthesis to produce an acceptable ear, and the green top could be used for fodder. Blackbirds were often pests, stealing the ripening kernels before they were mature, especially when the shuck did not completely cover the ear. To discourage the birds, as well as protect the grain from rain, farmers would often "break the stalk just below the ear with a blow of an iron bar, and then bend it downward." Before the final harvest in September or October, the best ears were selected to provide seed corn for the next year. A second crop was sometimes planted in June to provide sweet corn later in the summer, take advantage of labor freed up when the first crop was laid by, and counter the risk of losing the first planting.[13]

Anson Jones planted corn both for use on his plantation and to sell. Records from his Barrington plantation provide a good indication of how corn crops varied from year to year in planting and harvest dates, yields, and prices. In 1845 he was able to plant his crop by the end of January, and by the end of May he had plowed it three times. However, in 1846 bad weather forced him to replant some of his corn several times, and the corn had been plowed only twice by the end of May. Cold, wet weather also caused similar delays in 1850. In dry years such as 1851, when grass

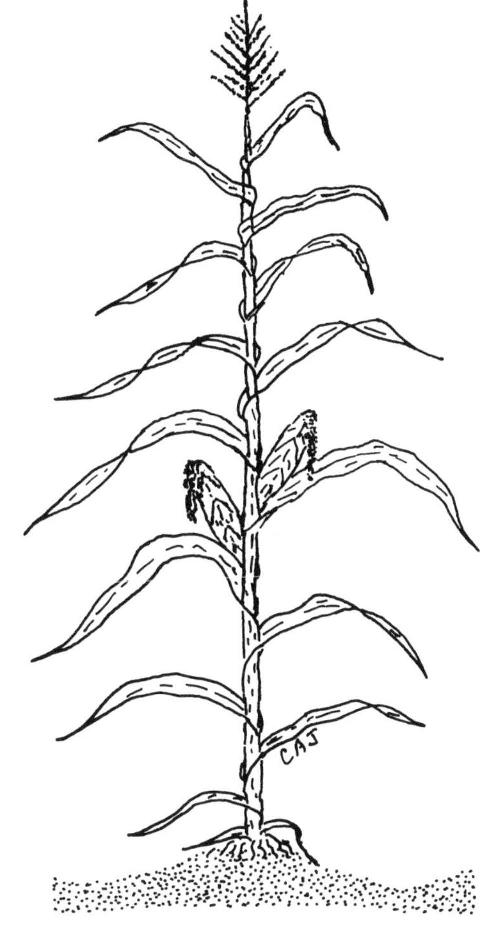

Corn plant

was scarce and the grain yields were expected to be poor, some of the corn crop was cut in July and fed as forage to livestock. The grain harvest typically began in mid-August. Because the mature corn could be left in the field without severe losses, the harvest could begin as early as mid-August, or it could be delayed until much later in the autumn. Prior to the Civil War, corn was usually harvested by hand and hauled by wagon to be stored until it was shucked. On plantations, corn was often shucked in wet weather when field work was not possible. In normal years, yields averaged from 40 to 80 bushels per acre; however, in the best years, newspapers boasted of yields of 90 to 110 bushels per acre. Though Jones fed most of the corn he produced to his livestock, he sometimes sold small amounts to neighbors. Between 1844 and 1853 he received from $0.40 to $1.00 per bushel for his corn. Of course, corn was typically cheaper immediately after harvest in the fall, when it often brought only two to four "bits" ($0.25 to $0.50) a bushel. In the spring it often rose to $1.00 a bushel, and in dry years such as 1855, prices of $1.50 to $2.50 were common. Political and economic problems also caused temporary spikes in corn prices, such as $5.00 a bushel after the Texas Revolution in 1836 and $3.50 a bushel during the nationwide financial panic of 1837.[14]

The most serious pest of corn grain was the weevil, which Affleck felt was "increasing in numbers, rather than diminishing." As usual, he had a great deal of advice about how to protect the crop. First, he judged that the hard-seeded "flint" varieties were less susceptible to the weevil than were the softer flour or "gourd-seed" types. He felt that it was unwise to leave corn "in the field until overrun with the weevils" and then to store it "in the shuck, by which all of the insects are stored with the grain." Therefore, the corn should be gathered and shucked "as soon as ripe," or it should be shucked "on the stalk, by which most of the parent insects would be left in the field." Finally, it was important to store the grain in clean, well-maintained cribs. Affleck recommended "neat, compact, frame corn-cribs, raised some feet from the ground; the space beneath being so arranged as to admit of being closed tight, when desired; the floors of the cribs slatted; the sides tight, with ventilators in the roof." They should be "carefully cleaned out, each year; fumigating with sulphur, thoroughly; and white-washing, adding a good share of salt to the wash." When a crib had been emptied, it should be cleaned out, fumigated by burning charcoal below the slatted floor, and whitewashed. The corn from another crib should then be transferred to the clean one, "with as few as possible of the weevil; destroying all that can be found."[15]

Though a small amount of corn was exported, Texians and their livestock consumed almost all they produced. The census of 1850 reported that in 1849 farmers produced 5,978,000 bushels of corn, an average of a little over 28 bushels for each of 212,592 Texians. By 1860 the population had grown to 604,215, but corn production increased apace, to 16,501,000 bushels—about 27 bushels per person.[16]

SUGARCANE

By the 1820s Louisiana had developed a substantial sugar industry, producing about 30,000 thousand-pound hogsheads in 1823 and about 45,000 in 1826. The success of the Louisiana industry stimulated the

Sugarcane mill

expansion of sugarcane production into adjacent southeastern Texas, where several plantations began to produce significant amounts of sugar in the late 1820s and early 1830s. Among the first Texas producers were Judge S. M. Williams on the Trinity River, Eli Mercer on the Colorado River in present-day Wharton County, and William Stafford on Oyster Creek in present-day Fort Bend County. In 1830 Williams "raised cane enough to make forty hogsheads of sugar of fine quality." Mercer crushed his cane between mills made of live oak stumps and turned by mules or oxen. Little effort was made to separate the sugar crystals from the molasses. As a result, early Texas sugar was described as "black as tar" and like "Mississippi alluvion, steeped in molasses." Nevertheless, these early efforts provided a valuable sweetener for the surrounding settlers, as well as a source of seed cane and experience that allowed for a rapid expansion of the industry after the Texas Revolution.[17]

Planters had to decide on which of the main cash crops, cotton or sugarcane, they would concentrate. In general, those with large slaveholdings often concentrated on sugar, whereas those with fewer slaves planted cotton. In 1836 the *Telegraph and Texas Register,* published in Columbia, recommended that "the farmer or planter without the resources . . . (say 50 hands) to engage in sugar making" could consider planting cotton "with only 5 to 20 hands." The newspaper reported that there were "several who successfully undertake this branch of agriculture [cotton production] with no other aid than the white individuals of their own family." However, those who preferred "a more easy mode of living," could "raise horses, mules, horned cattle, or hogs."[18]

In the early 1840s low cotton prices, wet weather, and increasing insect damage discouraged many cotton planters. As a result, some turned to sugarcane, and by 1842 one hundred hogsheads were produced on plantations along Caney Creek and the Colorado River. The first steam-powered mill was constructed by William Duncan on his Caney Creek plantation in 1843. In 1844 Duncan and John Sweeney, who farmed on the San Bernard, each produced one hundred hogsheads of sugar. Sweeney's sugar was described as far superior to the Louisiana sugar available in Texas. The sugar produced on William Menefee's plantation on the Colorado was pronounced "fully as light colored, dry and clean as the very best Louisiana sugar." Menefee's sugar could be bought for "about eight cents a pound, or about two cents less than the best article of Louisiana sugar on Texas markets."[19]

In the mid-1850s Olmsted observed several sugar plantations in the Guadalupe River bottom between Seguin and Victoria. Taking advantage of the warm coastal climate and an "abundance of fuel" to boil down the juice, planters grew cane "of unusual size, and perfectly developed." The land could be "bought, improved, for $10 per acre," and the planters were able to sell their sugar "at the highest price, for the supply of the back country." By 1847 there were at least five large sugar plantations on the Colorado and nine on the Brazos. Eli Mercer estimated that at least 1,800 hogsheads would be produced on the Brazos and 4,000 to 5,000 on Caney Creek and the Colorado. In addition, a large amount of seed cane was held back to expand acreages, and by 1849 Texas produced over 7,000 hogsheads, almost all in Brazoria and Matagorda counties. Though the growth of the industry was rapid, Texas lagged far behind Louisiana, which produced over 220,000 hogsheads in 1849.[20]

By the 1850s Brazoria County was among the most advanced agricultural counties in the state, producing more sugar than any other. I. T. Tinsley reported in the *Texas Almanac for 1859* that the county had three principal soil types: sandy "peach-land soils," alluvial "cane soils," and clayey "stiff black soils." Tinsley's two general rules for farming these flatland soils were that the land should be drained with ditches and "should be first ploughed up deep, and ploughed up well." On the stiff black soils, the main drainage ditches should be three to five feet deep and six feet wide at the top. Cross ditches about four feet wide at the top should be dug every seventy yards to drain water to the main ditch. Alluvial cane soils required less ditching than stiff black soils, with main ditches three to four feet deep, and light peach land needed even less.[21]

Production of sugarcane was a year-round activity, with planting taking place between December and March. Sugarcane was reproduced vegetatively using stalks, called "seed cane," from the previous crop. To plant new fields, the field hands cut the cane stalks in the fall. Because the cane was easily killed by freezing, the stalks were laid on the ground with their leafy tops turned to the south, covering stalks beneath, to provide protection from the north wind. The rows of cut cane were then covered with soil to provide additional protection and left until they were needed for planting. Tinsley recommended that cane lands should be "clear of trash, and well ridged up and pulverized with the harrow" before planting. The cane rows should be about six feet apart, and the furrow "should be deep and well opened." Cane stalks were stripped of their leaves and placed in a double row in the furrow. Col. J. D. Waters recommended that after dropping the cane in the furrow, "hands follow immediately . . . with sharp knives and cut each stalk in three pieces" to promote uniform growth of buds along the length of the canes. The cane was then covered from four to six inches deep with large turning plows and the ridge lightly harrowed and pressed down with a heavy roller to assure good contact between the cane and the soil.

By the end of February the planting was normally complete, and within a few weeks the danger of frost had passed. The soil over the rows was then scraped away, leaving the cane stalks two or three inches below the surface. As the soil warmed, the rows of cane shoots began to appear. If rains caused a crust to form on the soil surface, light harrows or hoes were used to help the cane shoots emerge. After the shoots emerged, slaves used hoes to chop any weeds that had begun to grow. After about six weeks a plow was run between the rows, throwing soil into the row to help the cane plants, or stools, put down more roots. Weeds were controlled by hoeing and plowing until late June, when the crop canopy was large enough to shade out weeds. Waters observed that "cane requires a higher cultivation than corn or cotton; in fact, to do it justice, it should be worked once in ten days, and very thoroughly each time."

From July to October the slaves prepared for the harvest and cane-grinding season. Hogsheads and molasses barrels were made, firewood was cut and hauled to the sugar house, and buildings and machinery were repaired. If the rains were adequate, the cane grew rapidly, forming erect, sugar-filled stalks up to eight feet long topped by a fanlike array of green leaves. Beginning in October the slaves used large, heavy cane knives to cut the stalks at ground level, remove the upper part of the stalk with its green leaves, and strip away any dead leaves that clung to the cane. The cutters placed the stalks of cane in piles on the ground until they were

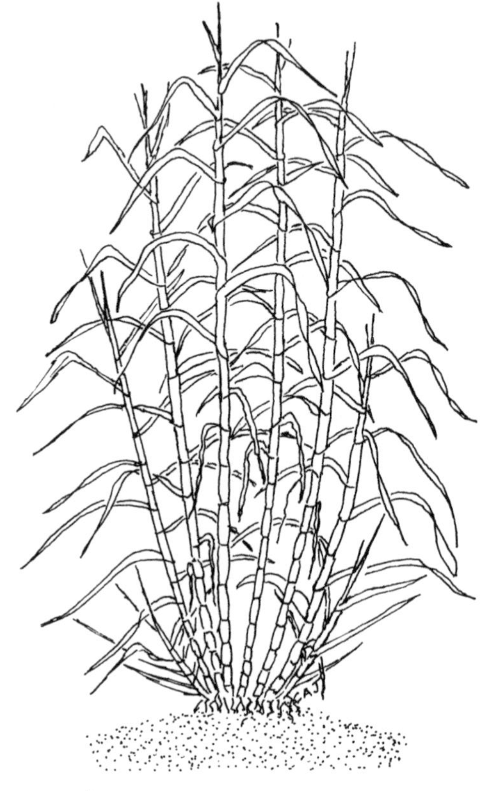

Sugarcane plant

picked up by another crew and carried in wagons or carts to the sugar house. After the crop was harvested, the cane plants began to regrow from buds located just below the soil surface. Though winter frosts often killed the young leaves, when the weather warmed in the spring, the "ratoon" crop would begin to grow, saving the planter the time and expense required to replant. Planters could often grow several ratoon crops before severe frost, insects, or drought reduced the stand enough to require replanting. An anonymous traveler reported that two varieties of sugarcane were grown. "The ribband [ribbon] cane requires to be planted every three years. The creole cane will continue to grow from the roots for ten or fifteen [years], and is but little inferior to it."[22]

THE SUGAR HOUSE

The freshly cut cane stalks were carried to the sugar house as soon as possible. There, the stalks were placed on a forty- or fifty-foot carrier made of chains with crosspieces of wood inserted in the links. As the carrier moved forward, it deposited the cane in a wooden hopper that guided it into the mill, which consisted of a pair of cylinders powered by a steam engine or, on some plantations, oxen. The mill's rotating cylinders crushed the cane, releasing its sweet juice into a large vat. The crushed cane stalks, known as bagasse, were conveyed down a trough to the outside, to be hauled away and discarded. The juice was strained to remove impurities, such as pieces of bagasse. It was then poured into a set of open kettles and heated to drive off water and concentrate the sugars that remained. A set of evaporation kettles consisted of four to six iron basins set in masonry over the horizontal flue of a wood-burning furnace. The kettles were arranged in order of descending size, with the fire under the smallest kettle. The hot gases from the fire passed under the kettles, heating the juice and causing it to evaporate.

The juice, mixed with slaked lime to neutralize its acidity and promote clarification, was first poured into the largest kettle. As the juice began to boil, it produced a foamy scum that was skimmed off the surface and discarded. The steam rising from the kettle left the sugar house through a steam chimney. As the juice became more concentrated, it was ladled into successively smaller kettles until it finally reached the concentration at which, when cooled, granules of sugar would form. The juice was then transferred into troughs to cool and crystallize. After six to fourteen hours, crystal formation was complete. The mixture of granulated sugar and molasses was then transferred to hogsheads placed above molasses cisterns, where they remained for twenty to thirty days while the molasses drained away, to be placed in barrels and sold. The raw sugar was then ready for shipment. Waters estimated that the average yield of sugar was sixteen hundred pounds per acre. "With fair management," each field hand could be expected to produce ten thousand pounds of sugar and eight hundred gallons of molasses per year.[23]

Developing a sugar plantation required substantial capital. In 1849 Morgan Smith imported sugar-making equipment worth thirty thousand dollars for his Waldeck plantation. It boasted a double set of kettles, vacuum pans to speed evaporation, centrifuges to separate the sugar crystals from molasses, and a brick sugar house resembling a "tesselated castle." His plantation, including mill, land, and more than one hundred slaves, was valued at $114,000 in 1853. A typical plantation with fifty slaves worth about six hundred dollars each was valued at approximately $50,000. But profits could be great. Smith produced 410 and 617 hogsheads in 1854 and 1855, respectively. Waters expected to make 300 hogsheads averaging fifteen hundred pounds each, grinding day and night and making 8 to 10 hogsheads each day. By 1850 Texas had more than thirty-five sugar plantations, with the majority using steam engines to drive their mills. In 1852 Texas produced its largest prewar sugar crop, more than 11,000 hogsheads. Forty-four plantations in four counties, Wharton, Matagorda, Fort Bend, and Brazoria, produced sugar, fifteen using horse-powered mills and twenty-nine with steam engines. Eight plantations produced more than 400 hogsheads each, and nineteen more made from 200 to 400 hogsheads. Brazoria was by far the most important county, with twenty-nine plantations producing over 8,200 hogsheads worth forty dollars each. This represented a gross return of $328,000 for the sugar plus $136,000 for molasses. The total value of these twenty-nine plantations was $1,134,000: $665,000 in slaves, $392,000 in buildings and equipment for the sugar houses, and $77,000 in land.[24]

After the banner year of 1852, a combination of low sugar prices and poor weather conditions discouraged planters. From a high of eight cents a pound in the mid-1840s, Texas sugar fell to four cents a pound (forty dollars for a thousand-pound hogshead) in 1852 and below three and a half cents in 1853. In 1854 a wet spring, a dry summer, and a September hurricane reduced production to about 7,500 hogsheads. But in 1855 both yields and prices recovered, with production exceeding 8,800 hogsheads and farmers receiving over nine cents a pound in January 1856. Unfortunately, the winter of 1855–56 was the most severe in years, with cold weather killing almost all the cane and wiping out the 1857 crop. Though the industry began to recover, only about 6,000 hogsheads were produced in 1858, and in 1859 the crop was about 5,100 hogsheads.[25]

In 1859 Brazoria County grew 7,287 acres of sugarcane; Matagorda and Fort Bend counties grew 1,160 acres and 690 acres, respectively. In addition, several counties far in the interior reported sugar production. Bell, Burleson, Burnet, Caldwell, Cherokee, Cooke, Dallas, Denton, DeWitt, Fannin, Gonzales, Hill, Jasper, Johnson, Liberty, Limestone, McLennan,

Navarro, and Washington each reported between 100 and 500 acres of sugar crops, nearly all of which was "Sorgo or Chinese Sugar [sweet sorghum] for home consumption."[26]

WHEAT

Wheat was introduced into "the prairies of Red River county by the first settlers, as early as 1833." However, prior to the 1850s, the remoteness of northern Texas from markets in Shreveport and Houston limited its wheat production to that needed for local consumption. In 1850 the total wheat production of Texas was reported to be about forty-two thousand bushels. With rapid migration into the northern counties during the 1850s, the production increased dramatically. J. W. Latimer, writing in the *Texas Almanac for 1859*, estimated that wheat production in the northern counties of Collin, Cook, Dallas, Denton, Ellis, Grayson, Johnson, Parker, Tarrant, and Wise exceeded two million bushels in 1858.

Even though wheat could be grown on a variety of soils in northern Texas, it was reportedly best adapted to "rich black prairie loam." One of the advantages of wheat was its ease of cultivation. Latimer reported that "the mass of wheat-growers in Texas, as yet, take but little pains and incur as little expense in its cultivation as possible." It was usually sown in September or October, after the corn crop was harvested. A few

Wheat heads

farmers plowed the field prior to planting with a grain drill. However, for most the only labor involved was "to fell the corn stalks, sow the seed broadcast . . . , and plough or harrow in." The "common red May" variety was used, and farmers "had but little interest in procuring and testing new varieties." When the crop was sown broadcast, one bushel of seed was planted per acre, and after the seed was "in the ground, nothing more [was] required until harvest." In total, growing a wheat crop in this way required only "one hand to twenty-five acres, only from one month to six weeks in the year—two or three weeks in planting, etc., and the same time in harvesting." If the fall-sown crop failed, a spring wheat variety could be sown in February or March. Latimer recommended grazing the wheat from December to mid-March, because it provided "the finest pasturage for stock," keeping cattle "sleek and fat through the most rigorous winter, until the rising of grass on the prairies, in the spring." In addition, grazing had an additional advantage. It delayed growth and emergence of the head, reducing the danger of frost damage in the spring.[27]

In 1851 Elise Waerenskjold contrasted wheat production in Norway and northeastern Texas. She wrote that "in Norway, if a person wants to turn a piece of unbroken land into a field, it will cost him much labor; even the cultivated soil must be fertilized if the yield is to be satisfactory—and even so the crop is often destroyed by frost." In contrast, on the Texas prairies "a person merely needs to fence and plow the land and it is ready to be sown." Echoing Latimer's observations, she reported that Texians generally "cultivated in a rather slipshod fashion." Farmers broadcast wheat seed "amidst the stalks and all the weeds" of the recently harvested corn crop, then plowed the fields and hoped for an adequate stand of wheat. After that, no further care was needed until harvest. Wheat was harvested in May "when the grain [was] in the transition from the 'dough' to a hard state." The wheat was usually cut with a scythe equipped with a "cradle" of thin wooden strips that supported the cut stems until they could be placed in a row on the ground. The stalks were tied in bundles and threshed by being "simply placed on the ground and trampled out by horses or oxen." Though much of the grain was lost, one of Waerenskjold's neighbors had "harvested thirty bushels per acre, having used four bushels of seed." Though his yield was "above average, to be sure," she felt they "could be much improved by more careful tillage."[28]

In the 1850s mechanization began to reach Texas. A Kentucky Harvester, requiring two or four horses and two laborers, could cut fifteen to twenty acres per day, and eight laborers were used to bind the

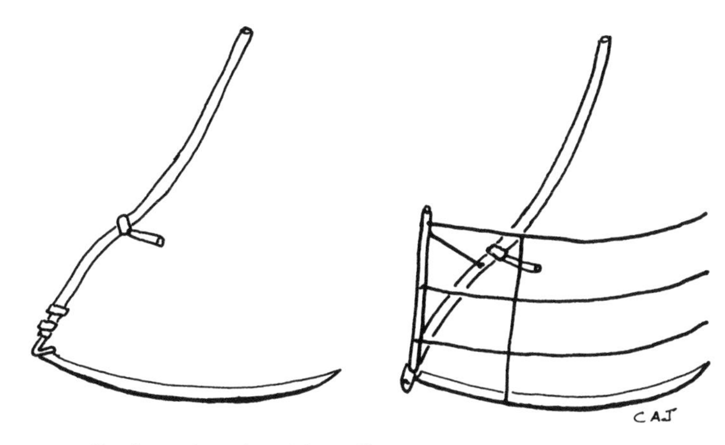

Scythe and scythe with cradle

straw into bundles and stack them in shocks for the grain to dry. After the wheat was dried, an "endless chain-thresher," requiring two horses and eight to ten laborers, was used to thresh and clean 150 to 300 bushels per day. Some farmers used Manny's or McCormick's reapers, and some did not bind and shock their wheat, allowing it to lie on the ground until dry enough to thresh. By 1858 Kilburn & Brotherton were manufacturing a machine in Dallas County "said to thresh and clean from 500 to 600 bushels per day." In addition, "the Messrs. Wilson, of the same county, [had] invented and patented valuable improvements to harvesters, one of which particularly, rakes the grain from the swath, and deposits it in the thresher at a great saving of labor." J. W. Latimer reported that the worst problems of wheat in the area were rust, "wheat-birds," and late frosts. In addition, grasshoppers sometimes attacked before frost in the fall, though the crop usually recovered in the spring.[29]

Farther south, in Williamson County, Samuel Mather wrote in 1858 that although "twenty bushels to the acre is just about a fair average [wheat] crop in Texas, taking the various kinds and qualities of soil . . . [and] seasons, good and bad, into the account . . . from thirty to forty bushels are doubtless very often made." Though scythes with cradles had been used to cut wheat in earlier years, two-horse "patent reapers" were "coming into general use." They could be purchased from L. G. Williams in Galveston for about $175, and two laborers using the reaper could cut about ten acres per day. Mather stated that "some four or five years ago, the threshing was all done by horses or by hand," but threshing machines "of various kinds" powered by horses, water, or steam were "now in almost universal use." Horse-powered machines could thresh from about 125 to 800 bushels per day, depending on the number of animals used for power. Mather recommended that as soon as the wheat crop was harvested, the land should immediately be plowed to a depth of about eight inches with a "common turning plough" to stimulate sprouting of weed and wheat seeds. These seedlings could then be turned under and killed before planting the next crop. Mather also indicated that "a few of the best farmers subsoil the land, loosening the ground to the depth of fourteen or fifteen inches, and this extra labor is always well paid for in the superior crop."[30]

Latimer estimated that the per-acre cost of wheat production in northern Texas was $3.00 for land rent, $1.00 for seed, $1.50 for sowing and plowing in the seed, $1.00 for reaping and binding, and $1.60 for threshing. At an average yield of twenty bushels per acre and an average price of $1.00 per bushel, the net profit was $11.90 per acre. However, in 1858 "the largely increased quantities cultivated the past season, the extended region in which it was grown, the aggregate heavy yield and consequent large surplus" had caused the price of wheat to drop precipitously, to $0.50 per bushel. The cost of hauling wheat or flour in ox wagons was $1.00 per hundred pounds per hundred miles. At that price, it was not feasible to ship wheat to markets in Shreveport, Galveston, or Houston. Latimer hoped that railroads would soon reach the region, reducing the cost of transportation and opening new markets for the crop. With characteristic optimism, the *Texas Almanac*'s editors predicted that in a few years they would "see trains of freight cars coming to Galveston heavily loaded with flour and wheat for shipment to Northern ports, to supply the Northern markets before their own wheat can be matured."[31]

THE FARMER'S TOOLS

Throughout the first half of the nineteenth century, land in the United States was cheap in comparison with the cost of labor needed to farm it. Farmers could often afford more land than they could farm, and many were eager to obtain labor-saving tools in order to farm more land. Inventors recognized their opportunity to tap into the expanding agricultural economy. As a result, the period from about 1820 to 1860 saw unprecedented invention and commercialization of labor-saving farm equipment. Austin's colonists, in Spanish Texas at the very beginning of this period, brought only the simplest tools with them, such as axes, hoes, scythes, and simple plows. The first plows used by the settlers were handmade and primitive, such as the "bull tongue shovel" plow described by

W. B. Allen. Made of wood that was covered and strengthened with "at least a dozen pieces of scrap-iron, old horse shoes and wagon bed irons," these plows could do little more than make a shallow furrow, killing weeds in their path but without turning the soil.[32]

One of the most important technological advances was in the design and manufacture of plows. In 1819 Jethro Wood of New York patented an improved cast-iron plow with interchangeable parts. These new plows used a share whose point penetrated below the soil surface, cutting and lifting a slice of soil several inches thick and almost a foot wide. As the plow moved forward, the slice of soil was pushed to the side by the share and turned completely over by the plow's curved moldboard, causing it to crumble and leaving a surface that could be harrowed and planted. But some prairie soils tended to stick to the early cast-iron moldboards rather than scour cleanly off it. In response to this problem, in 1837 John Deere invented a one-piece wrought-iron plow that scoured so well that it became known as the "singing plow." Deere's company soon improved its design, combining a steel share with a wrought-iron moldboard. By the late 1850s Deere was building ten thousand a year. Many other manufacturers began to produce plows prior to the Civil War, and Affleck's *Southern Rural Almanac for 1860* advertised "Plows, A Full Assortment of Sizes Made by James H. Hall of Maysville, Kentucky." Hall's sole agent in New Orleans was Samuel Lock. Hall cautioned planters to buy only his plow, branded "James H. Hall, Maysville, Ky." because "a spurious and inferior plow has been made in Pittsburgh, and sent to this market, branded on the beam 'Hall's Maysfield Pattern.'"[33]

As plows improved, farmers could prepare more land, requiring improvements in other farming activities, including sowing small grains, planting corn and cotton, controlling weeds, harvesting, and threshing small grains. Numerous inventors and manufacturers recognized this huge market for more efficient farm machinery, and from the 1830s to 1860 a large number of practical tools came onto the market. Machines needed to mechanize the sowing, harvesting, and threshing of wheat and other small grains were developed in the 1840s and 1850s. These included grain drills to sow the grain, reapers (including the famous McCormick reaper) to cut it, and threshers to remove the grain and separate it from the straw. In addition, threshers required power, which was provided by "horse powers." Two types of horse powers were used. The treadmill type used two horses walking in place on a moving belt to generate power that was transferred by a belt to the thresher. The sweep type used one or more horses walking in a circle to turn a horizontal geared wheel that transferred power through a rod along the ground or overhead to a set of gears and a belt that transferred power to the thresher. Corn planters, cultivators, and shellers came onto the market in the 1840s and 1850s, but mechanization of the corn harvest was delayed until after the Civil War. By 1860 Affleck's *Southern Rural Almanac* carried an advertisement for Peter & Buchanan's Seed and Agricultural Warehouse in Louisville, Kentucky. It offered seeds, steel plows, corn shellers, horse powers, threshers, and reaping and mowing machines.[34]

GRISTMILLS, FLOUR MILLS, AND SAWMILLS

As Texas grew, settlers needed to saw lumber and grind corn and wheat. Noah Smithwick noted that "old John Cummings, one of the Three Hundred, made his usefulness in the colony manifest by building the first mill in the state on the creek to which he gave his name, a few miles above San Felipe." It was a sawmill "with a corn cracker attached, all run by water." By 1826 Texas had several gristmills (corn mills) operated by hand or animal power. A typical animal-powered mill consisted of a horizontal beam with each end fastened about ten feet above the ground to a tree. In the middle of the beam was a hole into which a vertical shaft was inserted, extending downward to the millstones. Horses or mules, hitched to a horizontal arm attached to the shaft, walked in a circle to turn the top millstone, which ground the corn against the lower stationary stone. In 1831 William Rabb, another of Austin's Old Three Hundred, constructed a gristmill on the east side of the Colorado River in present-day Fayette County. Though he was able to purchase some of the machinery for the mill in New Orleans, his millstones, or burrs, were imported from Scotland. When they arrived in Matagorda, Rabb was unable to ship them up the river because of logjams. Therefore, he simply attached the two millstones to the ends of an axle and, with the stones as wheels, used oxen to roll them to the mill. In 1833 Stephen F. Austin reported that his colony had two steam- and six water-powered mills, powering either gristmills, saws, or both.[35]

John Lockhart, one of the early settlers in Washington County, noted that prior to the establishment of sawmills, boards were made "with the broad-axe and fro [frow]" and were "rough and uncouth." The earliest sawmills used a sash saw, a single vertical blade in a frame that moved up and down, cutting one board at a time. Powered by animals, water, or steam engines, sash saws could cut from five hundred to twelve hundred board-feet a day. The first steam-powered saws began to operate in the early 1830s. In the 1840s the first sawmill was built in Washington

County. Powered by the waters of New Year's Creek, the mill used a sash saw to rip logs from "a fine cedar-break" into rough planks, "which were left in what was called a feather-edge finish." Carpenters then trimmed and squared the edges with drawknives and planes to make finished boards. Lockhart recalled that settlers came from long distances to purchase the boards, which they were glad to get "at any price" in order to cover the walls of their log houses and "to build neat frame houses, and put on much style—as it was called." Lockhart recalled that the first "revolving saw that came into use" had "an iron bar with a chisel-shaped cutter on each end." The new revolving saws were called "wood-pecker" saws "from the pecky way in which they operated."[36]

Many of the water-powered mills were designed both to saw lumber and to grind corn and wheat. In 1850 a community of Mormons built a mill on Hamilton Creek about eight miles below Burnet. In addition to grinding grain, the mill powered a sawmill and lathes, and the community manufactured furniture. However, the colony became indebted, and in 1853 the mill was sold to Noah Smithwick. The three-story wooden frame mill stood at the base of "a precipice twenty-eight feet or more in height" over which Hamilton Creek flowed, falling into a deep pool below. A flume conducted a portion of the creek's flow from the top of the falls to "the huge overshot wheel, twenty-six feet in diameter, which furnished power for the mill." Smithwick reported that "great clumsy, rattling wooden cog wheels and drum and fly-wheels filled up the lower stories, the upper one containing a small corn cracker mill and an old up-and-down sash saw" that could handle larger logs than could the circular saws then available. Dissatisfied with the mill's performance, Smithwick build a new overshot wheel and replaced much of the machinery, "throwing out all the old wooden cog work and substituting [cast-iron] casings." He then installed "a new set of burrs and added bolting works, the first flouring mill west of Georgetown." The availability of a mill capable of producing wheat flour "gave a new direction to the farming interest" in the region, and "soon the rattle of the threshing machine was heard in the land, and the reign of the corn-dodger was over in those parts." People came from miles around and sometimes had to wait for several days to have their wheat ground into flour.[37]

Hubbs Brothers of New Orleans sold the Coleman's Farm Mill, a steel mill described as "perfectly simple in construction." An advertisement in Affleck's *Southern Rural Almanac for 1860* claimed that the mill could grind eight to fifteen bushels of corn per hour. It could be powered by "one, two, three, or four horse[s], or by steam or water power, or by hand." It could be used to grind "corn or cob, or any kind of grain for feed, hominy of any grade, or the finest family meal, and superfine flour with equal facility, bolting or sifting while grinding." The price of the mill was seventy-five dollars when equipped with a sieve to sift cornmeal. For an extra twenty-five dollars, the mill could be purchased with an attached bolting machine to sift wheat flour.[38]

The use of steam power increased rapidly after 1845, and by 1860 some two hundred Texas sawmills had twelve hundred employees. Steam engines and circular saws could be purchased in New Orleans. Stationary steam engines from one and a half to thirty horsepower and circular sawmills with blades up to seventy-two inches in diameter could be purchased from Samuel H. Gillman of Gravier Street, New Orleans. The Baronne Street Steam Boiler Manufactory, also in New Orleans, made boilers to power steam engines. With more steam engines to drive mills and steamboats, cordwood became increasingly important. F. W. C. Cook of New Orleans was the sole agent for the "Planters' Patented Cord Wood Cutter." The machine consisted of a boiler mounted on a pair of iron wheels that could be pulled by a yoke of oxen "while the steam is up ready for work." Steam was conveyed to one or two saws up to two hundred feet away through a "flexible composition steam pipe." These reciprocating saws were five and a half to eight feet long and weighed 150 to 200 pounds. Using the Cord Wood Cutter, two men with one saw could cut forty to sixty cords of wood per day. The boiler "with slight alteration," could also be used to drive threshing machines, shingle machines, corn mills, cotton gins, straw cutters, or pumps.[39]

TRANSPORTATION

Though Texas' fertile soils could grow a wide variety of crops, the cost of transportation, as much as any other problem, limited the expansion of agriculture up the rivers and onto the fertile Texas prairies. The most efficient way to transport cotton and other agricultural products to market was by river. Large river-bottom plantations built landings and loaded cotton onto rafts, flatboats, keelboats, or steamboats for the trip down the river. However, when the river was high, heavily loaded rafts could overturn. Rafts and boats often ran aground on sandbars when the water was low, and sunken logs could snag them. The mouth of the Brazos was long blocked by a sandbar, and logjams, also called rafts, blocked the Guadalupe and Colorado rivers until the 1840s. For many years a massive raft on the Red River above Shreveport hindered transportation through Caddo Lake to Jefferson.[40]

If a plantation was not on a navigable stream, its crop was hauled to the river in "cotton wagons, two or three sometimes together, drawn by three or four pairs of mules or oxen_._._. each managed by its Negro driver." Before the arrival of flatboats and keelboats, log rafts were poled down the river with their load of bales. In the early 1830s Robert Hunter, his father, and two other men tied fifty-nine bales of cotton onto "2 large canoes maid [*sic*] out of 2 large cotton wood trees, with a platform on." They started off on Christmas morning, but the raft soon hit a snag and overturned. The cold, wet men cut the cotton loose from the raft and floated it to shore. They then borrowed a yoke of oxen and pulled the bales up the wet bank. The next spring the Hunters built another raft, rolled the bales back down to the river, floated them to Columbia, and sold the bales, but for only three dollars a piece.[41]

Flatboats were only slightly more elaborate. Like rafts, they consisted of logs tied together, but they had a wooden roof to cover the bales. Both rafts and flatboats were steered with poles, and their timbers were sold for lumber when they reached their destination. Keelboats were true boats, pointed at each end with a hold for the cotton and a cabin with windows on the deck. Steered by a large rudder, keelboats were floated down the river. But they could also make the return trip, either poled by slaves or pulled by lines hitched to oxen plodding along the bank, usually loaded with goods for the plantation or for sale by small-town merchants. Several keelboats operated on the Colorado River in the late 1830s and 1840s. One, the *Davy Crockett,* drew three feet of water with a cargo of six hundred bales of cotton. In 1843 William Bollaert reported seeing a keelboat being built to carry two hundred to three hundred bales of cotton from La Grange down the Colorado River to the raft of logs blocking its mouth. Launched at Matagorda in 1845, the keelboat *Kate Ward* was 110 feet long with a 24-foot beam.[42]

Steamboats offered great advantages, both for marketing Texas cotton and for bringing manufactured goods back up the river to the plantations and merchants. With their large cargo capacity and speed, they were enthusiastically welcome when they began to appear in about 1830. The *Branch T. Archer, Ellen Frankland, Vesta,* and *Scioto Belle* plied the Trinity in the late 1830s and 1840s. Regular steamboat service on Buffalo Bayou between Galveston and Houston began by 1840. Boats such as the *Cayuga, Yellow Stone, Washington,* and *Brazos* played important roles in development of agriculture and towns along the Brazos. A ball was held when the *Cayuga* arrived in San Felipe in 1833. The *Yellow Stone* helped carry Sam Houston's army across the Brazos ahead of Santa Anna and then raced a load of cotton down the river under fire from the

Mexican Army. Washington, later known as Washington-on-the-Brazos, was up the river from San Felipe near the confluence of the Brazos and Navasota rivers. An important distribution point for agricultural supplies as early as 1835, it became even more important when the steamer *Mustang* arrived in 1842. This boat once reached Port Sullivan, near Hearne. Jefferson became an important shipping point for cotton from northeastern Texas after the steamboat *Llama* found a way through the Red River raft and across Caddo Lake into Big Cypress Creek in late 1843 or early 1844.

But by 1844 the Texas economy was so bad and so many steamboats had been wrecked that only one, the *Dayton,* remained in service. Keelboats and flatboats were again pressed into service to move cotton down the rivers. After Texas joined the Union, steamboats were brought back into service on Texas rivers. In 1848 the *Texian Advocate,* a newspaper published in Victoria, reported steamboats on the Guadalupe, the Brazos, the Trinity, and the Neches. In 1848 residents formed the Brazos Steamship Association to eliminate obstacles to navigation, and the association's two steamboats, the *Washington* and the *Brazos,* began operating regularly between Washington and Velasco, near the mouth of the Brazos. The *Washington* even made its way up the Brazos and the Little River to Cameron the winter of 1850–51, though during normal flows the practical limit of navigation was Washington-on-the-Brazos.

La Grange was the normal limit of navigation on the Colorado, though Olmsted reported that "the smallest class of boats" could sometimes ascend as far as Austin. In 1849 the *Texas Republican* advised its readers that William Perry of Jefferson had been contracted for a fee of thirty-five hundred dollars to remove obstructions from Caddo Lake. Steamboats could normally reach Cincinnati (near Huntsville) on the Trinity. Other normal limits of navigation were Victoria on the Guadalupe, Bevilport (near Jasper) on the Neches, Gaines Ferry (near San Augustine) on the Sabine, and Marion on the Angelina. Steamboats also made their way for thirty miles up Caney Creek and up Buffalo Bayou to Harrisburg. Despite their advantages over land transportation, steamboats were notoriously unsafe. Boiler explosions and fires were not unusual, and Texas newspapers provided details of both the disasters and the attendant losses of merchandise.[43]

Most of the cotton from northeastern Texas was carried by wagon to Jefferson, where it was loaded onto steamers that made their way down Big Cypress Creek, across Caddo Lake, and down the Red River to New Orleans. The *Texas Almanac for 1860* estimated that in 1859 at least two hundred thousand bales of Texas cotton from the northeastern portion of the state had been shipped down the Red River for sale in New Orleans. Over half of this total was loaded at Jefferson.[44]

Ports were also established on the bays of the Texas coast. Seagoing vessels unloaded manufactured goods and lumber, then loaded cotton, sugar, hides, pelts, tallow, and other agricultural goods. The most important port on the eastern coast was Galveston. To the southwest, Indianola and Port Lavaca competed for freight entering and leaving Matagorda Bay. Numerous lesser ports and river landings were also established, sometimes receiving oceangoing ships, but often used to load freight onto small boats for transfer to the major ports.[45]

Some Texian planters and stock raisers, anxious to improve their livestock, bought breeding stock from the United States and even England. Breeding stock was routinely driven from the eastern United States. In 1840 the English ship *Ironsides* brought improved breeds of hogs, sheep, cattle, and horses to Texas; in 1850 the *Palmetto* brought breeding stock.[46]

Cotton was by far the most important export from Galveston, increasing from 65,809 bales worth $2,701,500 in 1853 to 148,362 bales valued at $8,139,910 in 1859. The largest cargoes were from 3,000 to 4,000 bales of cotton, and in 1859 a total of 495 vessels with an aggregate tonnage of 281,734 arrived in port. During the 1850s the percentage of Galveston's exports sent to U.S. ports, especially New Orleans, declined, and by 1859 about two-thirds of Galveston's cotton exports were shipped directly to Great Britain.[47]

INDIANOLA VERSUS PORT LAVACA

Reports of port activity give us a good sense of the kinds and amounts of goods exported and imported. The *Indianola Courier* reported that during the year ending August 1, 1859, imports to Indianola totaled over 3.1 million feet of lumber and 343,533 barrels (each containing five cubic feet) of manufactured goods. In addition, salt, fruit, and "a number of carriages" were among the imports. Exports included 28,622 bales of cotton, 1,675 bales of wool, 1,602 cattle, 33,109 hides, 26,508 bushels of pecans, 350 hogsheads of sugar, 838 barrels of molasses, 150 bales of peltries, 14,150 bushels of corn, "several thousand" pounds of bacon, 28 barrels of Texas wine,

28 barrels of butter, 62 barrels of sea island cotton, 325 sheep, and "a large number" of horses and mules. In addition to these agricultural products, 992 slabs of lead, 33,550 pounds of copper, and 1,126 packages of "sundries" were exported.[48]

J. H. Davis made a separate report for imports passing through Port Lavaca in the year ending August 1, 1859. They included 228,209 "Barrels of five Cubic feet" (presumably of manufactured goods), more than 2.2 million feet of lumber, 208,000 laths, more than 1.2 million shingles, and 250,000 railroad ties. For the same period, agricultural exports included 30,440 bales of cotton, 1,204 bags of wool, 34,887 hides, 509 bales of peltries, 10,973 horns, 38 casks of bacon, 5,882 bushels of corn, 106 barrels of "sour" flour, 1,909,252 pounds of pecans, 912 beef cattle, 12 horses, and 3 barrels of "Mustang Grape Wine." Other exports included 40 boxes of merchandise, 20 tons of "old Iron," and 36 bales of moss. Exports reported to be of Mexican origin included 176,488 pounds of lead, 78,587 pounds of copper, and 142,505 dollars "in Bullion and Mexican dollars." Sailing ships visiting Port Lavaca during the year included seven barks, four brigs, and ninety-three schooners, "chiefly from New York." In addition, the port was visited by thirty-four steamships, "exclusive of the two lines of the Southern Steamship Company, from New Orleans."[49]

It appears that, in keeping with the competition between Indianola and Port Lavaca, both ports reported many of the same exports and imports. The *Texas Almanac* did not attempt to arbitrate the dispute, noting only that "the [Lavaca] Herald says a number of the vessels reported by the [Indianola] Courier for Indianola, were consigned to Lavaca and discharged there." The *Herald* contended that "five barrels out of every six imported from the North to Matagorda Bay go up to Lavaca . . . [and] of all the cotton shipped from Matagorda Bay, amounting to about 35,000 bales, 30,444 bales were received and shipped from Lavaca."[50]

As the number of planters increased, plantations were established on the prairies farther from navigable rivers and seaports. John Sneed, owner of a plantation with sixty slaves eight miles south of Austin, hauled his cotton, corn, and wheat to Port Lavaca, near Indianola. His trains consisted of twenty-five or thirty wagons, each pulled by four or five oxen. The routes taken to reach seaports were sometimes circuitous. For example, John Duncan, who had a plantation on Caney Creek, hauled cotton westward, crossing the Colorado River at Elliott's Ferry. From there the cotton continued overland to Wilson Creek, where it was transferred to flatboats that had been poled up the creek to receive it. The cotton was then taken down the creek to Tres Palacios Creek, Tres Palacios Bay, and on to Palacios Point. On their return trips planters brought a variety of goods needed on the plantation.[51]

John Lockhart boasted that even though the portion of Washington County near Chappell Hill was not situated on a major river, during the 1840s it "prospered more rapidly in commerce and agriculture than any part of middle Texas." Because river transportation was not available, teamsters hauled cotton to Houston on ox wagons at a cost of fifty cents per hundred pounds. One yoke of oxen was normally required for each bale on the wagon. In good weather the round trip to Houston required two weeks; in rainy weather with muddy roads the same trip took almost a month. Lockhart recalled that "all the interior of Texas," from Dallas, Corsicana, and Waco to Austin, Bastrop, and La Grange, depended on teamsters for transportation of cotton and other freight. Each teamster had "a complete camping outfit, and always took with him a Spanish pony to ride when driving his oxen." Time was no object to the teamster. He stopped "an hour or two before dark and hobbled his oxen and pony" to allow them to graze. His supper consisted of "fried bacon, hoe-cake and black coffee," and he slept in a tent or under the stars in a bedroll. In the morning "he made no haste, for his cattle, as he called his oxen, must have time to graze, as it is generally known cattle do not as a usual thing graze much of the night." After saddling his pony, he drove the oxen back to the wagon, yoked them, and began the day's drive by 10 or 10:30 in the morning. Like most Texans, the teamsters "were never known to walk when it could possibly be avoided. . . . They would walk a half-mile" in order to catch a horse "to ride a quarter[-mile]."[52]

When travelers needed to spend the night on the road, they often chose to camp out, especially when they could find good pastures for their horses and the weather was good. However, in rainy or cold weather a room and a bed were greatly appreciated. By the 1840s many farmhouses, especially in eastern Texas, were "considered as houses of 'entertainment,' or inns." They charged $1.00 to $1.50 "for supper, bed, and breakfast and stabling for the horse; but the majority of farmers [would] not receive payment . . . [and] under no circumstances [was] a traveller rich or poor turned from the door."[53]

During wet weather wagons and carts often became stuck in deep mud holes, and the rivers were frequently so high that they could not be crossed. During the winter of 1843 an English couple, the E. M. Houstouns, attempted to travel from Houston to Washington-on-the-Brazos, which rainy weather prevented them from reaching. One stretch of two miles "was not traversed in less time than four hours, so deep was the mire." Mrs. Houstoun reported that "travelers are seen knee-deep in mud, and looking as though hopeless of rescue, and dying and dead cattle are interspersed among bales of cotton, which are in process of 'hauling.'" In addition, she noted that "the Brazos and Trinity bottoms are overflowed for weeks together in the winter season, and, in the absence of causeways and bridges, are extremely difficult and even dangerous to pass."[54]

Olmsted reported that in 1854 cotton was hauled on wagons, "from all parts of the state, to Houston, Indianola, or Red River, unless its owners [were] content to leave it an indefinite period upon the nearest river-bank, subject to the vague chances of a rise." Planters in eastern Texas sent their cotton, usually on wagons loaded with five bales of about four hundred pounds each, to the Sabine. If the Sabine was too low, as it was in 1854 when Olmsted visited, the crop was hauled "over execrable roads, distances of 100 and even 150 miles" to Natchitoches or Grand Ecore in Louisiana, where the Red River joined the Mississippi. The cost of transportation to the river was about one cent per pound for each fifty miles traveled. River transportation on the Sabine to Galveston or from Grand Ecore to New Orleans cost about one to one and a fourth cents a pound.[55]

Counties realized that the value of their lands and their economic growth depended on moving goods to market. Because the state had no agency responsible for road construction and maintenance, counties took the lead by dividing themselves into road districts with overseers responsible for assuring that residents contributed labor to road maintenance. Slave owners normally contributed bondsmen, whereas those who were not slaveholders usually provided family labor.[56]

From 1845 to 1860 the firm of Howard and Ogden was well known, using mule teams to haul freight for the U.S. Army from Indianola to San Antonio, then transferring it to oxcarts for movement to frontier forts. At one time this firm alone had eight hundred teams. By 1852 three to four thousand ox wagons, each with sixteen to twenty-four oxen, were providing freight service from Houston to Austin, Waco, and all of eastern Texas. Up to a thousand teams were reported in Jefferson at one time bringing cotton for shipment by steamboats to New Orleans, at a normal cost of one cent per mile per hundred pounds.[57]

In 1855 a newspaper reported that in the previous season thirty-eight thousand bales of cotton were received in Houston from distances of twenty to two hundred miles away. This required an estimated 1,566 wagons costing $150 each and twenty-five thousand yokes of oxen costing $50 each. On average, each wagon carried six bales per trip and made four trips per season at a cost of $100 per trip. By the Civil War at least ten thousand teams were operating throughout the state, and they were a strong political force resisting the development of railroads.[58]

By the late 1850s stage lines provided rapid transportation for passengers and light freight between most Texas communities. Each coach was drawn by a team of four or six horses or mules at an average speed of six to ten miles an hour. The *Texas Almanac for 1859* reported that at a "usual price" of about ten cents a mile, "coach and hack services" carrying the U.S. mail would carry passengers between most towns in Texas. In 1860 travelers could leave Galveston at 1 P.M. on "the fine, staunch, fast- running Steamer *Swan*" and arrive a Liberty at 4 A.M. the next morning. At 8 A.M. Sawyer's Line of four-horse coaches left Liberty and proceeded via Smithfield, Livingston, Moscow, Sumpter, Colita, Pine Point, and Shady Grove to Crockett, where it arrived the following day at noon. Another route from Galveston to Crockett was "by steamboat to Houston; thence by Central and Houston Railroad to Cypress; thence by four-horse coaches, via Rosehill, Montgomery, Huntsville and Cincinnati." From Houston, the traveler could also go by rail to Hempstead, then by coach to Waco, Nacogdoches, or Austin. Four-horse coaches also carried passengers from Nacogdoches to Waco, Brenham, Austin, and Shreveport and Alexandria, Louisiana. The Great Northern mail ran between Clarksville and San Antonio via New Braunfels, San Marcos, Austin, Georgetown, Belton, Waco, Mellford, Chambers' Creek, Waxahachie, Lancaster, Dallas, McKinney, Bonham, and Paris. By the time the Civil War began, a network of thirty-one stage lines allowed travelers, often with several transfers, to reach almost any significant town in Texas.[59]

When Texas gained its independence from Mexico in 1836, the entire United States had only 1,273 miles of railroads. Texas had none. The difficulty and cost of transporting cotton and other freight by oxcart were powerful incentives for construction of railroads. Anxious to bring this symbol of modernity to the new nation, the first charter to build a railroad in Texas was issued to the Texas Railroad, Navigation, and Banking Company in December 1836. Over the course of the next twenty-five years, sixty more charters were issued; however, when Texas joined the Confederacy in 1861, only ten railroads were in operation, with a total of 468 miles of track. The most important Texas railroads built before the Civil War reached out toward the richest and most populated region of the state—the Brazos valley. The first track for the Buffalo Bayou, Brazos, and Colorado Railroad (BBB&C) was laid in 1852, crossing the Brazos at Richmond in Fort Bend County. In 1856 planters in Brazoria County financed the Houston Tap and Brazoria Railroad (HT&B), which ran south from the BBB&C south to Columbia, then west to Wharton on the Colorado. The Houston and Texas Central Railroad (H&TC) ran from Houston to Hempstead in Austin County, then northward through Grimes County. The Washington County Railroad ran from the H&TC at Hempstead to Chappell Hill and Brenham. The planters' willingness to promote and finance these railroads is understandable. They not only reduced the cost of marketing the region's cotton and sugar but their presence raised land values.[60]

Cotton, corn, sugarcane, and wheat were the principal crops in antebellum Texas. All benefited from the state's fertile soils. Cotton production grew rapidly as farmers migrated to Texas and began to produce for the relatively stable European and northeastern markets prior to the Civil War. Corn was the most important food crop of both farm families and the slaves on Texas plantations. As a result, it enjoyed good local markets as both human food and livestock feed. Though Texas sugar production grew rapidly through 1852, it was only a small fraction of the much larger Louisiana industry. Beginning in the mid-1850s a combination of low prices, drought, hurricanes, and frosts began to reduce enthusiasm for the crop. Wheat production in the northern parts of the blackland prairies and in the rolling hills west of Austin and San Antonio held great promise. More easily grown than corn, it was well suited for the small farmers of these regions. Gradually, animal-, water-, and steam-powered mills were built, reducing the labor associated with grinding corn and wheat, squeezing sugarcane, and sawing lumber. As it became possible to produce more of these commodities, farmers and planters were faced with the greater challenge of moving them to distant markets. Water was the preferred and most economical means of transport to markets such as Galveston and New Orleans. But as cotton and wheat production expanded onto the fertile prairie soils and river valleys of the central and northern parts of the state, ox-drawn wagons provided the necessary transportation to market.

Texians, if not Tejanos, had reason to be optimistic as the 1850s drew to a close. The republic had been absorbed into the Union, and the Mexican War had finally ended most hostilities toward its southern neighbor. The U.S. Army and tens of thousands of settlers had pushed the frontier well beyond San Antonio and Austin, and Indian depredations had declined substantially. But hundreds of Tejano families had been intimidated or enticed into selling their ranches, usually to Texians and at a fraction of their value. Tejano residents had been driven out of several towns, and Texian teamsters had, through murder and intimidation, displaced their Tejano competitors between San Antonio and the coast. Despite these setbacks for most Tejanos, a few wealthy families were able to adapt to the new political situation. Some land-rich Tejanas married Texians or recently discharged veterans of the Mexican War, and together the couples expanded their ranching empires.

In addition, the state's registration of land titles was working. Thousands of Spanish, Mexican, and Republic of Texas land grants had been patented, and cheap, fertile land was attracting hundreds of thousands of immigrants, many with substantial resources. As a result, the state's population and agricultural production had almost tripled during the decade. Planters had taken advantage of the booming economy. The number of slaves had more than tripled, and the number of planters producing at least forty bales of cotton or forty hogsheads of sugar had grown more than sixfold since 1850, to over twenty-four hundred. Thousands of ox teams carried farm products to market, and stagecoach lines connected all the major towns. By the 1850s the state's first railroads had begun to reach the major agricultural areas along the lower Brazos and Colorado rivers. The nationwide boom in agricultural mechanization had begun to reach Texas, promising larger crops with less labor in the future.

However, the forces leading to secession and civil war had been gathering for years. Antipathy between the northern and southern states was growing, and firebrands on both sides were tired of compromise. After Abraham Lincoln's election in November 1860, and despite Governor Sam Houston's steadfast opposition, the Texas legislature met in special session, called a convention that voted overwhelmingly for secession, held a referendum that approved secession, and Texas formally joined the Confederacy on March 5, 1861. These decisions would radically change Texas agriculture, as well as every other aspect of its society and economy.

NOTES

CHAPTER 1

1. Teresa Rojas Rabiela, "La agricultura en la época prehispánica," in *La agricultura en tierras mexicanas desde sus orígenes hasta nuestros días*, ed. Teresa Rojas Rabiela, 15–138.

2. Rojas Rabiela, "Época prehispánica," 15–138; María de los Ángeles Romero Frizzi, "La agricultura en la época colonial," in *La agricultura en tierras mexicanas desde sus orígenes hasta nuestros días*, ed. Teresa Rojas Rabiela, 139–215.

3. Rojas Rabiela, "Época prehispánica," 15–138; William E. Doolittle, *Canal Irrigation in Prehistoric Mexico*.

4. Romero Frizzi, "Época colonial," 139–215.

5. Romero Frizzi, "Época colonial," 139–215; François Chevalier, *Land and Society in Colonial Mexico*, trans. Alvin Eustis, 50–65.

6. Romero Frizzi, "Época colonial," 139–215; Chevalier, *Land and Society*, 50–65.

7. Chevalier, *Land and Society*, 66–69.

8. Charles J. Bishko, "The Peninsular Background of Latin American Cattle Ranching," *Hispanic American Historical Review* 32 (November 1952): 491–515; David Dary, *Cowboy Culture: A Saga of Five Centuries*, 11–12; Paul H. Carlson, *Texas Woollybacks*, 10–11.

9. Terry G. Jordan, *North American Cattle-Ranching Frontiers*, 123–45; Joe S. Graham, *El Rancho in South Texas: Continuity and Change from 1750*, 12.

10. Donald E. Chipman, *Spanish Texas, 1519–1821*, 43–55; Jordan, *Cattle-Ranching Frontiers*, 123–45; Chevalier, *Land and Society*, 92–109.

11. Chipman, *Spanish Texas*, 43–55; Jordan, *Cattle-Ranching Frontiers*, 123–45; Chevalier, *Land and Society*, 92–96.

12. Chevalier, *Land and Society*, 96–98, 104–8; Jesús de la Teja, "A Fine Country with Broad Plains—the Most Beautiful in New Spain," in *On the Border: An Environmental History of San Antonio*, ed. Char Miller, 41–55.

13. Dary, *Cowboy Culture*, 13–15; Graham, *El Rancho in South Texas*, 13–15.

14. Chevalier, *Land and Society*, 112–14; Dary, *Cowboy Culture*, 18–21; Graham, *El Rancho in South Texas*, 13–15.

15. Don Worcester, *The Spanish Mustang*, 6–13, 22–23.

16. Chipman, *Spanish Texas*, 70–85.

17. Chipman, *Spanish Texas*, 86–99.

18. Jack Jackson, *Los mesteños*, 10; Chipman, *Spanish Texas*, 70–85, 99–116.

19. David La Vere, *The Texas Indians*, 103–12; Fray Isidro Felis de Espinosa, "Descriptions of the Tejas or Asinai Indians, 1691–1722. Part IV. Fray Isidro Felis de Espinosa on the Asinai and Their Allies," trans. Mattie Austin Hatcher, *Southwestern Historical Quarterly* 30 (October 1927): 154; Henri Joutel, "Historical Journal," in *The Journeys of René Robert Cavelier, Sieur de La Salle*, ed. Isaac Joslin Cox, 2:137–38.

20. John R. Swanton, *Source Material on the History and Ethnology of the Caddo Indians*, 127–30.

21. Espinosa, "Descriptions of the Tejas," 156; Joutel, "Historical Journal," 138.

22. Swanton, *Source Material*, 127–30.

23. Espinosa, "Descriptions of the Tejas," 156–57; Richard I. Ford, "Preliminary Report on the Plant Remains from the George C. Davis Site, Cherokee County, Texas, 1968–1970 Excavations," App. 1 in "1968–1970 Archeological Investigations at the George C. Davis Site, Cherokee County, Texas," by Dee Ann Story, *Bulletin of the Texas Archeological Society* 68 (1997): 104–7.

24. Espinosa, "Descriptions of the Tejas," 152, 156–57.

25. Espinosa, "Descriptions of the Tejas," 153; Ross C. Fields, Eloise F. Gadus, and L. Wayne Klement, "The Peerless Bottoms Site: A Late Caddoan Component at Cooper Lake, Hopkins County, Texas," *Bulletin of the Texas Archeological Society* 65 (1994): 55–114; Ford, "Preliminary Report"; Mark Walters and Patti Haskins, "Archaeological Investigations at the Redwine Site (41SM193), Smith County, Texas," *Journal of Northeast Texas Archaeology* 11 (1998): 1–38.

26. Billy Davidson, "Preliminary Study of the Faunal Remains from the George C. Davis Site, 1968–70 Excavations," App. 2 in "1968–1970 Archeological Investigations at the George C. Davis Site, Cherokee County, Texas," by Dee Ann Story, *Bulletin of the Texas Archeological Society* 68 (1997): 108–13; Espinosa, "Descriptions of the Tejas," 153; Timothy K. Perttula et al., "The Carlisle Site (41WD46), a Middle Caddoan Occupation on the Sabine River, Wood County, Texas," *Notes on Northeast Texas Archaeology* 1 (1993): 34–62; Walters and Haskins, "Redwine Site," 1–38.

27. Fray Francisco Casañas de Jesús Maria, "Descriptions of the Tejas or Asinai Indians, 1691–1722. Part I. Fray Francisco Casañas de Jesús Maria to the Viceroy of Mexico," trans. Mattie Austin Hatcher, *Southwestern Historical Quarterly* 30 (January 1927): 211; The Gentleman of Elvas, "The Narrative of the Expedition of Hernando de Soto," in *Spanish Explorers in the Southern United States, 1528–1543*, ed. Theodore H. Lewis, 220; Espinosa, "Descriptions of the Tejas," 157.

28. Dan L. Flores, *Jefferson and Southwestern Exploration*, 169–70.

29. Father Anastasius Douay, "Narrative of La Salle's Attempt to Ascend the Mississippi in 1687," in *The Journeys of René Robert Cavelier, Sieur de La Salle*, ed. Isaac J. Cox, 1:232–34; Espinosa, "Descriptions of the Tejas," 179.

30. Espinosa, "Descriptions of the Tejas," 157; Ohland Morton, *Terán and Texas: A Chapter in Texas-Mexican Relations*, 68, cited in Jean Louis Berlandier, *The Indians of Texas in 1830*, ed. John C. Ewers, trans. Patricia Reading Leclercq, 47; Timothy K. Perttula, *The Caddo Nation: Archaeological and Ethnohistoric Perspectives*, 209.

31. Espinosa, "Descriptions of the Tejas," 156–57.

32. David La Vere, *Life among the Texas Indians: The WPA Narratives*, 83; Vynola Beaver Newkumet and Howard L. Meredith, *Hasinai: A Traditional History of the Caddo Confederacy*, 31–32; Espinosa, "Descriptions of the Tejas," 172; Pierre Margry, ed., *Découvertes et établissements des français dans l'ouest et dans le sud de l'Amérique Septentrionale, 1614–1754*, 3:367; Pierre Marie François de Pagés, *Travels round the World in the Years 1767, 1768, 1769, 1770, 1771*, 61.

33. Casañas, "Descriptions of the Tejas," 211; Espinosa, "Descriptions of the Tejas," 152; Margry, *Découvertes*, 3:394–95; Newkumet and Meredith, *Hasinai*, 32–33.

34. Fields, Gadus, and Klement, "Peerless Bottoms Site"; Ford, "Preliminary Report," 55–114; Claude McCrocklin, 1999, personal communication; La Vere, *Life among the Texas Indians*, 82–84; Walters and Haskins, "Redwine Site," 1–38.

35. La Vere, *Life among the Texas Indians*, 84; Casañas, "Descriptions of the Tejas," 212; Joutel, "Historical Journal," 143.

36. Mildred S. Gleason, *Caddo: A Survey of the Caddo Indians in Northeast Texas and Marion County—1541–1840*, 35–38; La Vere, *Life among the Texas Indians*, 91–92; Timothy K. Perttula, "The Archeology of the Pineywoods and Post Oak Savanna of Northeast Texas," *Bulletin of the Texas Archeological Society* 66 (1995): 331–59; Timothy K. Perttula et al., "Prehistoric and Historic Aboriginal Ceramics in Texas," *Bulletin of the Texas Archeological Society* 66 (1995): 175–235; La Vere, *Life among the Texas Indians*, 92.

37. Joutel, "Historical Journal," 138–39; F. H. Douglas, "The Grass House of the Wichita and Caddo," Denver Art Museum, Department of Indian Art, Leaflet no. 42 (February 1932), reprinted in *The Southern Caddo: An Anthology*, ed. H. F. Gregory.

CHAPTER 2

1. Donald Chipman, *Spanish Texas, 1519–1821*, 117.

2. Andrew Forest Muir, ed., *Texas in 1837, an Anonymous Contemporary Narrative*, 98–99.

3. Chipman, *Spanish Texas*, 119–27.

4. Thomas F. Glick, *The Old World Background of the Irrigation System of San Antonio, Texas*, 41; Mardith K. Schuetz, *Excavation of a Section of the Acequia Madre in Bexar County, Texas and Archeological Investigations at Mission San José in April 1968*, 3–6; Fray Juan Agustín Morfí, *History of Texas: 1673–1779*, trans. Carlos E. Castañeda, 225; Jesús de la Teja, *San Antonio de Béxar: A Community on New Spain's Northern Frontier*, 32, 75–80; Edwin P. Arneson, "Early Irrigation in Texas," *Southwestern Historical Quarterly* 25, no. 2 (October 1921): 121–30; Betty Eakle Dobkins, *The Spanish Element in Texas Water Law*, 108–13.

5. Chipman, *Spanish Texas,* 129–31; Jack Jackson, *Los mesteños,* 36–39; Robert H. Thonhoff, *The Texas Connection with the American Revolution,* 12–14.

6. Arneson, "Early Irrigation in Texas," 121–30; Dobkins, *Texas Water Law,* 108–13.

7. Arneson, "Early Irrigation in Texas," 121–30; Wells A. Hutchins, "The Community Acequia: Its Origin and Development," *Southwestern Historical Quarterly* 31 (1927–28): 28; Michael C. Meyer, *Water in the Hispanic Southwest,* 44; Schuetz, *Acequia Madre,* 3–6; Dobkins, *Texas Water Law,* 108–13.

8. Jesús F. de la Teja and John Wheat, "Béxar: Profile of a Tejano Community, 1820–1832," in *Tejano Origins in Eighteenth Century San Antonio,* ed. Gerald E. Poyo and Gilberto M. Hinojosa, 8–9; Chipman, *Spanish Texas,* 133; William Edward Dunn, "Apache Relations in Texas, 1718–1750," *Quarterly of the Texas State Historical Association* 14 (1911): 198–274; Elizabeth A. John, *Storms Brewed in Other Men's Worlds,* 263–74.

9. Jackson, *Los mesteños,* 16–17; Chipman, *Spanish Texas,* 138; Muir, *Texas in 1837,* 108, 110.

10. Carlos E. Castañeda, *Our Catholic Heritage in Texas, 1519–1936,* 2:298.

11. Béxar Archives Translations, 2:19–237; Arneson, "Early Irrigation in Texas," 121–30; Teja, *San Antonio de Béxar,* 74–80; Dobkins, *Texas Water Law,* 108–13.

12. Arneson, "Early Irrigation in Texas," 121–30; Dobkins, *Texas Water Law,* 108–13.

13. Béxar Archives, 10:7–198; 18:58–153.

14. Robert S. Weddle and Robert H. Thonhoff, *Drama and Conflict: The Texas Saga of 1776,* 16–19.

15. Béxar Archives, 2:19–237.

16. Ibid., 6:83–167.

17. Ibid., 9:14–30.

18. Jackson, *Los mesteños,* 16–17, 20–21; Béxar Archives, 23:153, 30:3–4; Sandra L. Myers, *The Ranch in Spanish Texas: 1691–1800,* 36; Béxar Archives, 30:3–4.

19. Béxar Archives, 37:27–37; Arneson, "Early Irrigation in Texas," 121–30; Dobkins, *Texas Water Law,* 116–18.

20. Castañeda, *Our Catholic Heritage,* 2:208; Herbert E. Bolton, *Texas in the Middle Eighteenth Century: Studies in Spanish Colonial History and Administration,* 19; Jackson, *Los mesteños,* 34.

21. Teja, *San Antonio de Béxar,* 92.

22. María de los Ángeles Romero Frizzi, "La agricultura en la época colonial," in *La agricultura en tierras mexicanas desde sus orígenes hasta nuestros días,* ed. Teresa Rojas Rabiela, 139–215.

23. Castañeda, *Our Catholic Heritage,* 2:205–10.

24. Béxar Archives, 11:9–11.

25. John, *Storms Brewed,* 274–85; Jackson, *Los mesteños,* 22.

26. V. W. Lehmann, *Forgotten Legions: Sheep in the Rio Grande Plain of Texas,* 10–15; Castañeda, *Our Catholic Heritage,* 4:5–8, 9–11, 13–15, 36, 265–66.

27. Jackson, *Los mesteños,* 36–37; Castañeda, *Our Catholic Heritage,* 4:11–14; Weddle and Thonhoff, *Drama and Conflict,* 153.

28. Castañeda, *Our Catholic Heritage,* 4:5–6.

29. Bolton, *Texas in the Middle Eighteenth Century,* 97–99.

30. Bolton, *Texas in the Middle Eighteenth Century,* 97–99; Pierre Marie Françoiś de Pagés, *Travels round the World in the Years 1767, 1768, 1769, 1770, 1771,* 101–2.

31. Bolton, *Texas in the Middle Eighteenth Century*, 99–100; Jackson, *Los mesteños*, 36–37; Thonhoff, *Texas Connection*, 12–14; James Wakefield Burke, *Missions of Old Texas*, 132.

32. Pagés, *Travels round the World*, 56, 72–73, 77–78.

33. Peter P. Forrestal and Paul J. Foik, trans. and eds., *The Solís Diary of 1767*, vol. 1, no. 6, *Preliminary Studies of the Texas Catholic Historical Society*, 8–9.

34. Jackson, *Los mesteños*, 133–34; Jean Louis Berlandier, *Journey to Mexico during the Years 1826 to 1834*, trans. Sheila M. Ohlendorf, Josette M. Bigelow, and Mary M. Standifer, 1:271–74.

35. Béxar Archives, 4:1–68; 22:66–170.

CHAPTER 3

1. Jack Jackson, *Los mesteños*, 51–57; Robert S. Weddle and Robert H. Thonhoff, *Drama and Conflict: The Texas Saga of 1776*, 145–46.

2. Béxar Archives Translations, 30:54–60.

3. Jackson, *Los mesteños*, 60–63; Robert H. Thonhoff, *The Texas Connection with the American Revolution*, 15–17; Robert H. Thonhoff, *El Fuerte del Cíbolo*, 54; Weddle and Thonhoff, *Drama and Conflict*, 144–59.

4. Pierre Marie François de Pagés, *Travels round the World in the Years 1767, 1768, 1769, 1770, 1771*, 56, 82–83, 96–99; Sandra L. Myers, *The Ranch in Spanish Texas*, 25.

5. Pagés, *Travels round the World*, 97.

6. Jackson, *Los mesteños*, 9–10.

7. Donald E. Chipman, *Spanish Texas, 1519–1821*, 183–86; Weddle and Thonhoff, *Drama and Conflict*, 59–61; Jackson, *Los mesteños*, 125–47.

8. Béxar Archives, 64:1–187; Jackson, *Los mesteños*, 133–47.

9. Béxar Archives, 68:3–13.

10. Fray Juan Agustín de Morphí, *Excerpts from the Memorias for the History of the Province of Texas*, 58–59.

11. Béxar Archives, 68:3–13.

12. Jackson, *Los mesteños*, 618–21; Meyers, *Ranch in Spanish Texas*, 17, 37; Weddle and Thonhoff, *Drama and Conflict*, 46; Béxar Archives, 68:58–66.

13. Jackson, *Los mesteños*, 191–92.

14. Béxar Archives, 72:1–140; 73:1–105; 74:1–94; 75:53–137; 76:22–25, 37–38.

15. Jackson, *Los mesteños*, 62, 185–87; Jesús F. de la Teja, *San Antonio de Béxar: A Community on New Spain's Northern Frontier*, 114–15.

16. Myers, *Ranch in Spanish Texas*, 26–27; Jackson, *Los mesteños*, 35–50; Terry G. Jordan, *North American Cattle-Ranching Frontiers*, 147–56; Thonhoff, *El Fuerte del Cíbolo*, 63.

17. Jackson, *Los mesteños*, 228–31; Weddle and Thonhoff, *Drama and Conflict*, 154.

18. Myers, *Ranch in Spanish Texas*, 47–49; Weddle and Thonhoff, *Drama and Conflict*, 46.

19. Jackson, *Los mesteños*, 620–21.

20. Béxar Archives, 130:18–20; 131:62–63.

21. Ibid., 127:13–15.

22. Ibid., 131:62–63; 126:6–59.

23. Ibid., 127:45–113.

24. J. Autrey Dabbs, trans., *The Texas Missions in 1785*, vol. 3, no. 6, *Preliminary Studies of the Texas Catholic Historical Society*, 12–14.
25. Teja, *San Antonio de Béxar*, 109; Jackson, *Los mesteños*, 396.
26. Teja, *San Antonio de Béxar*, 106–7.

CHAPTER 4

1. Florence Johnson Scott, *Royal Land Grants North of the Río Grande*, 6–7; Edwin J. Foscue, "Agricultural History of the Lower Río Grande Valley Region," *Agricultural History* 8: 124–37.
2. Scott, *Royal Land Grants*, 7–9; Foscue, "Agricultural History," 124–37.
3. Scott, *Royal Land Grants*, 9–12; Jack Jackson, *Los mesteños*, 443.
4. Scott, *Royal Land Grants*, 12–15; Robert S. Weddle and Robert H. Thonhoff, *Drama and Conflict: The Texas Saga of 1776*, 162; Foscue, "Agricultural History," 124–37; Betty Eakle Dobkins, *The Spanish Element in Texas Water Law*, 128–30.
5. Foscue, "Agricultural History," 124–37; Texas General Land Office, *Guide to Spanish and Mexican Land Grants in South Texas*; Jackson, *Los mesteños*, 444–50.
6. Armando C. Alonzo, *Tejano Legacy*, 80–83.
7. Scott, *Royal Land Grants*, 60–61.
8. Berlandier, *Journey to Mexico*, 1:262–66, 2:423.
9. Ibid., 1:262–64.
10. Ibid., 2:426–38.
11. Alonzo, *Tejano Legacy*, 90–91.
12. Teresa Griffin Vielé, *Following the Drum*, 157, 160, 130–34.
13. Alonzo, *Tejano Legacy*, 79–80; Berlandier, *Journey to Mexico*, 2:430, 542–44.
14. Andrés Tijerina, *Tejano Empire: Life on the South Texas Ranchos*, 3–11.
15. Ibid., 11–15.
16. Terry G. Jordan, *North American Cattle-Ranching Frontiers*, 123–38, 147–58.
17. Tijerina, *Tejano Empire*, 5–6; Joe S. Graham, *El Rancho in South Texas: Continuity and Change from 1750*, 25.
18. Tijerina, *Tejano Empire*, 21–28, 34–35; Graham, *El Rancho in South Texas*, 24.
19. Tijerina, *Tejano Empire*, 23–25.
20. Tijerina, *Tejano Empire*, 32–33; Graham, *El Rancho in South Texas*, 25.
21. Graham, *El Rancho in South Texas*, 23–24; Andrés Sáenz, *Early Tejano Ranching: Daily Life at Ranchos San José and El Fresnillo*, 10–11, 42–43; Tijerina, *Tejano Empire*, 36–37.
22. Noah Smithwick, *The Evolution of a State or Recollections of Old Texas Days*, 9–10.
23. Tijerina, *Tejano Empire*, 39–44; Béxar Archives Translations, 4:1–68 and 22:66–170.
24. Jackson, *Los mesteños*, 79–84; Pierre Marie François de Pagés, *Travels round the World in the Years 1767, 1768, 1769, 1770, 1771*, 54–56.
25. Jackson, *Los mesteños*, 79–85, 88; Zebulon Montgomery Pike, *The Journals of Zebulon Montgomery Pike*, 2:88.
26. Jackson, *Los mesteños*, 81; Joe S. Graham, "*Vaquero* Folk Arts and Crafts in South Texas," in *Hecho en Tejas*, ed. Joe S. Graham, 93–116; Graham, *El Rancho in South Texas*, 29; David Dary, *Cowboy Culture: A Saga of Five Centuries*, 31.
27. Mary S. Helm, *Scraps of Early Texas History*, ed. and ann. Lorraine Jeter, 92–93.
28. Pike, *Journals of Zebulon Montgomery Pike*, 2:77–78.

29. Berlandier, *Journey to Mexico*, 2:546.

30. Don Worcester, *The Spanish Mustang*, 26–27; John Q. Anderson, *Tales of Frontier Texas, 1830–1860*, 17.

31. Andrew Forest Muir, ed., *Texas in 1837, an Anonymous Contemporary Narrative*, 109–10.

CHAPTER 5

1. Béxar Archives Translations, 36:169–78.

2. Béxar Archives, 39:6–7; Pierre Marie François de Pagés, *Travels round the World in the Years 1767, 1768, 1769, 1770, 1771*, 88–89, 91–92, 98–99, 101; Béxar Archives, 46:65–67.

3. Béxar Archives, 48:129, 136; 49:1–11.

4. Ibid., 56:15–18, 37–43, 58–59.

5. Béxar Archives, 63:62; Robert S. Weddle and Robert H. Thonhoff, *Drama and Conflict: The Texas Saga of 1776*, 26–38.

6. Robert H. Thonhoff, *El Fuerte del Cíbolo*, 63.

7. Béxar Archives, 126:111–15.

8. Ibid., 127:36–41.

9. Donald E. Chipman, *Spanish Texas, 1519–1821*, 198–99.

10. Béxar Archives, 17:7–8; 20:338–40; 26:47–49; 30:1–2; 62:16–19.

11. Béxar Archives, 48:161–65; 53:32–36; 65:120; 66:45; Weddle and Thonhoff, *Drama and Conflict*, 16–19.

12. James Christopher Harrison, "The Failure of Spain in East Texas: The Occupation and Abandonment of Nacogdoches, 1779–1821" (Ph.D. diss., University of Nebraska–Lincoln, 1980), 368–73.

13. Béxar Archives, 60:12–15.

14. Ibid., 55:17–87; 57:1–50.

15. Ibid., 60:58–107.

16. Sandra L. Myers, *The Ranch in Spanish Texas: 1691–1800*, 28, 44–47.

17. Zebulon Montgomery Pike, *The Journals of Zebulon Montgomery Pike*, 2:73–74; Béxar Archives, 48:155–58; 56:67–116; 61:22.

18. Carlos E. Castañeda, *Our Catholic Heritage in Texas, 1519–1936*, 2:209.

19. María Esther Domínguez, *San Antonio, Tejas, en la época colonial (1718–1821)*, 277; Benedict Leutenegger, ed. and trans., *Inventory of the Mission San Antonio de Valero: 1772*, 25–38.

20. Domínguez, *San Antonio, Tejas*, 277; Leutenegger, *Inventory of Valero*, 25–38.

21. Benedict Leutenegger, ed. and trans., *Inventory of Mission Purissima Concepcion in the Province of Texas Made When the Fathers of Santa Cruz Gave It Up in 1772*, Archives of the College of Zacatecas, Old Spanish Missions Research Library, microfilm roll 4.5634-4.5664, 14–30.

22. Fray Juan Agustín Morfí, *History of Texas: 1673–1779*, trans. Carlos E. Castañeda, 95–98.

23. Domínguez, *San Antonio, Tejas*, 277; Morfí, *History of Texas*, 99.

24. Fray Marion A. Habig, *The Alamo Chain of Missions: A History of San Antonio's Five Old Missions*, 84–85; Morfi, *History of Texas*, 99; J. Autrey Dabbs, trans., *The Texas Missions in 1785*, vol. 3, no. 6, *Preliminary Studies of the Texas Catholic Historical Society*, 6–10.

25. Béxar Archives, 67:112–113; 79:3–14; 81:84–86.
26. Morfí, *History of Texas*, 92–93.
27. Jesús F. de la Teja, *San Antonio de Béxar: A Community on New Spain's Northern Frontier*, 120–22.
28. V. W. Lehmann, *Forgotten Legions: Sheep in the Rio Grande Plain of Texas*, 18–21; M. A. Hatcher, *The Opening of Texas to Foreign Settlement: 1801–1821*, Bull. no. 2714, 303–5, 357–58.
29. James Wakefield Burke, *Missions of Old Texas*, 127; Morfí, *History of Texas*, 100–1; Béxar Archives, 75:35–42; Weddle and Thonhoff, *Drama and Conflict*, 157.
30. Castañeda, *Our Catholic Heritage*, 5:40; Chipman, *Spanish Texas*, 205; Thomas F. Glick, *The Old World Background of the Irrigation System of San Antonio, Texas*, 41.
31. Jackson, *Los mesteños*, 487–88.
32. Ibid., 507–11.
33. Jackson, *Los mesteños*, 511–23, app. D, H, I, J.
34. Jackson, *Los mesteños*, 444–50, 490–523, app. D, G, I; Andrés Tijerina, *Tejanos and Texas under the Mexican Flag, 1821–1836*, 17–24.
35. Chipman, *Spanish Texas*, 216–41; Tijerina, *Tejanos and Texas*, 13, 19.
36. Tijerina, *Tejanos and Texas*, 69–90.
37. Jean Louis Berlandier, *Journey to Mexico during the Years 1826 to 1834*, trans. Sheila M. Ohlendorf, Josette M. Bigelow, and Mary M. Standifer, 2:289–300.
38. Berlandier, *Journey to Mexico*, 2:290–99.
39. Noah Smithwick, *The Evolution of a State or Recollections of Old Texas Days*, 31–34.
40. Berlandier, *Journey to Mexico*, 2:289–300; Jesús F. de la Teja and John Wheat, "Béxar: Profile of a Tejano Community, 1820–1832," in Tejano Origins in Eighteenth Century San Antonio, ed. Gerald E. Poyo and Gilberto M. Hinojosa, 2–19; Andrew Forest Muir, ed., *Texas in 1837, an Anonymous Contemporary Narrative*, 110–11; Ferdinand Roemer, *Texas, with Particular Reference to German Immigration and the Physical Appearance of the Country*, trans. Oswald Mueller, 120, 127.
41. Berlandier, *Journey to Mexico*, 2:550–51, 374–75; Tijerina, *Tejanos and Texas*, 12, 50.
42. Berlandier, *Journey to Mexico*, 2:553–56; Jackson, *Los mesteños*, 512–20.
43. Armando C. Alonzo, *Tejano Legacy*, 80–83.

CHAPTER 6

1. Paul W. Gates, *The Farmer's Age: Agriculture 1815–1860*, vol. 3, *The Economic History of the United States*, 1–21; T. R. Fehrenbach, *Lone Star: A History of Texas and the Texans*, 110–31.
2. Curtis Bishop, *Lots of Land*, 17–27; Fehrenbach, *Lone Star*, 132–38.
3. Fehrenbach, *Lone Star*, 132–40.
4. Ibid., 142–47.
5. Noah Smithwick, *The Evolution of a State or Recollections of Old Texas Days*, 39–70.

6. J. C. Clopper, "J. C. Clopper's Journal and Book of Memoranda for 1828," *Quarterly of the Texas State Historical Association* 13 (July 1909): 44–80.

7. Eugene C. Barker, ed., *History of Texas*, 118–19, 128, 133.

8. Mary S. Helm, *Scraps of Early Texas History*, ed. and ann. Lorraine Jeter, 46–49, 90–91.

9. Barker, *History of Texas*, 142; *A Visit to Texas*, 192–93; John C. Duval, *Early Times in Texas; or, The Adventures of Jack Dobell*, 231–36.

10. Clopper, "Clopper's Journal," 44–80; Felix Robertson, "Dr. Felix Robertson's Report," *The Weekly Messenger* (Russellville, Kentucky), June 17, 1826, 2, and *The Village Messenger* (Fayetteville), Tennessee, July 5, 1826, 2; W. B. Dewees, *Letters from an Early Settler of Texas*, 248; Abigail Curlee, "A Study of Texas Slave Plantations, 1822–1865" (Ph.D. diss., University of Texas, 1932), 199.

11. Elise Waerenskjold, *The Lady with the Pen*, ed. C. A. Clausen, 36–37.

12. Robert Hancock Hunter, *The Narrative of Robert Hancock Hunter*, 8.

13. Seymour V. Connor, "Log Cabins in Texas," *Southwestern Historical Quarterly* 53: 105–16; Terry G. Jordan, *Texas Log Buildings: A Folk Architecture*, 31–81.

14. Connor, "Log Cabins in Texas," 105–16; Jordan, *Texas Log Buildings*, 49–81.

15. Connor, "Log Cabins in Texas," 105–16; Taylor J. Allen, *Early Pioneer Days in Texas*, 3; Jordan, *Texas Log Buildings*, 90–94.

16. Allen, *Early Pioneer Days*, 3, 75–76; Connor, "Log Cabins in Texas," 105–16; Jordan, *Texas Log Buildings*, 83–103.

17. Allen, *Early Pioneer Days*, 4; Connor, "Log Cabins in Texas," 105–16; Smithwick, *Evolution of a State*, 89; Jordan, *Texas Log Buildings*, 83–103.

18. Allen, *Early Pioneer Days*, 4, 15–16, 76.

19. J. W. Wilbarger, *Indian Depredations in Texas*, ix–xii, 254–55; Barker, *History of Texas*, 345–58; William C. Pool, *A Historical Atlas of Texas*, 76–78, 108–15.

20. Daniel Shipman, *Frontier Life*, 55–56, 36.

21. Allen, *Early Pioneer Days*, 17.

22. Ferdinand Roemer, *Texas, with Particular Reference to German Immigration and the Physical Appearance of the Country*, trans. Oswald Mueller, 144–45; Frederick Law Olmsted, *A Journey through Texas*, 136–37, 266; Terry G. Jordan, *German Seed in Texas Soil: Immigrant Farmers in Nineteenth-Century Texas*, 164–65.

23. Thomas Affleck, *Affleck's Southern Rural Almanac, and Plantation and Garden Calendar for 1860*, 104–5, 114–15; Curlee, "Study of Texas Slave Plantations," 334–35; Sue Winton Moss, "Construction and Development of the Barrington Plantation," in *The Anson Jones Plantation: Archaeological and Historical Investigations at 41WT5 and 41WT6, Washington County, Texas*, 49–65.

24. F. M. Cross, *A Short Sketch-History from Personal Reminiscences of Early Days in Central Texas*, 3–5.

25. Olmsted, *Journey through Texas*, 88, 359.

26. Ibid., 100–2.

27. Hunter, *Narrative of Robert Hancock Hunter*, 8.

28. William Ransom Hogan, *The Texas Republic: A Social and Economic History*, 32–33; Smithwick, *Evolution of a State*, 8, 14.

29. Hogan, *Texas Republic*, 33–45.

30. Sue Winton Moss, "A Plantation Model for Texas," in *The Anson Jones Plantation: Archaeological and Historical Investigations at 41WT5 and 41WT6, Washington County, Texas*, 67–85; Hogan, *Texas Republic*, 33–38; Andrew Forest Muir, ed., *Texas in 1837, an Anonymous Contemporary Narrative*, 74.

31. Waerenskjold, *Lady with the Pen*, 29, 37.

32. Ibid., 36.

33. Smithwick, *Evolution of a State*, 8; Roger W. Rodgers, "Candid Columns: Life as Revealed in Antebellum Newspaper Advertising in Northeast Texas," *East Texas Historical Association* 38, no. 2 (2000): 40–53; Fehrenbach, *Lone Star*, 253–58; Hogan, *Texas Republic*, 81–109; Jonnie Lockhart Wallis, *Sixty Years on the Brazos: The Life and Letters of Dr. John Washington Lockhart, 1824–1900*, 197.

34. Sean Michael Kelley, "Plantation Frontiers: Race, Ethnicity, and Family along the Brazos River of Texas, 1821–1886" (Ph.D. diss., University of Texas at Austin, 2000), 71–75.

35. Hogan, *Texas Republic*, 81–93; Smithwick, *Evolution of a State*, 63.

36. Bishop, *Lots of Land*, 113–49; Dorman H. Winfre, *Julien Sidney Devereux and his Monte Verdi Plantation*, 32–91; *Texas Almanac for 1857*, "Report of November 1855," 37–38.

37. Andrés Tijerina, *Tejanos and Texas under the Mexican Flag, 1821–1836*, 137–40; Montejano, *Anglos and Mexicans*, 26–34; Arnoldo De León, *The Tejano Community, 1836–1900*, 14–15.

38. Alonzo, *Tejano Legacy*, 88–89; Scott, *Royal Land Grants*, 21–78; Jackson, *Los Mesteños*, 445, 638–43; Foscue, "Agricultural History"; David Montejano, *Anglos and Mexicans in the Making of Texas, 1836–1986*, 30–34; De León, *Tejano Community*, 14–15; Randolph B. Campbell, *Gone to Texas: A History of the Lone Star State*, 192.

39. Jordan, *German Seed*, 40–54.

40. Kelley, "Plantation Frontiers," 227–28; Jordan, *German Seed*, 55–112.

41. Olmsted, *Journey through Texas*, 140–41.

42. Ibid., 234–35.

43. Cat Spring Agricultural Society, *Century of Agricultural Progress: 1856–1956*, 10–16.

44. Olmsted, *Journey through Texas*, 194–96.

45. Ibid., 278–81.

46. Jordan, *German Seed*, 60–191.

47. Olmsted, *Journey through Texas*, 205–9, 459–60.

48. Ibid., 415–16.

49. Muir, *Texas in 1837*, 128–29; Archie P. McDonald, *Hurrah for Texas! The Diary of Adolphus Sterne, 1838–1851*, 23–24, 28, 36, 49; Abigail Curlee, "The History of a Texas Slave Plantation, 1831–63," *Southwestern Historical Quarterly* 26 (October 1922): 79–127; Tijerina, *Tejanos and Texas*, 12, 48–50; James M. Coleman, *Aesculapius on the Colorado*, 26; Marie Beth Jones, *Peach Point Plantation: The First 150 Years*, 34–35.

50. Affleck, *Southern Rural Almanac*, 52, 61–62, 38, 71–72; Coleman, *Aesculapius on the Colorado*, 28.

51. Dorman H. Winfrey, *Julien Sidney Devereux and His Monte Verdi Plantation*, 84–88.

52. Coleman, *Aesculapius on the Colorado*, 29, 42.

CHAPTER 7

1. Lester G. Bugbee, "Slavery in Early Texas," *Political Science Quarterly* 13 (September 1898): 389–412.

2. Sean Michael Kelley, "Plantation Frontiers: Race, Ethnicity, and Family along the Brazos River of Texas, 1821–1886" (Ph.D. diss., University of Texas at Austin, 2000), 81, 401; *Texas Almanac for 1858*, "African Slavery," 132–33.

3. Frederick Law Olmsted, *A Journey through Texas*, 62.

4. Ibid., 55–56.

5. Olmsted, *Journey through Texas*, 474; Richard G. Lowe and Randolph B. Campbell, *Planters and Plain Folk: Agriculture in Antebellum Texas*, 29–33; Kelley, "Plantation Frontiers," 402.

6. Randolph B. Campbell, *An Empire for Slavery: The Peculiar Institution in Texas, 1821–1865*, 50–58; Lowe and Campbell, *Planters and Plain Folk*, 29–33.

7. Lowe and Campbell, *Planters and Plain Folk*, 43, 46–55.

8. Abigail Curlee, "A Study of Texas Slave Plantations, 1822–1865" (Ph.D. diss., University of Texas, 1932), 218; Lowe and Campbell, *Planters and Plain Folk*, 120–21, 153, 188, 191; Campbell, *Empire for Slavery*, 50–58.

9. Curlee, "Study of Texas Slave Plantations," 219–21.

10. Lowe and Campbell, *Planters and Plain Folk*, 40–42, 59–74.

11. Mary S. Helm, *Scraps of Early Texas History*, ed. and ann. Lorraine Jeter, 96–98; Curlee, "Study of Texas Slave Plantations," 206.

12. Olmsted, *Journey through Texas*, 133–34; Samuel Mather, "Remarks on the Cultivation of Wheat in Williamson and Adjoining Counties," in *Texas Almanac for 1859*, 74–76; Thomas Affleck, *Affleck's Southern Rural Almanac, and Plantation and Garden Calendar for 1860*, 20, 89.

13. Dewees, *Letters* 248–50, 299–300.

14. Abigail Curlee Holbrook, "Cotton Marketing in Antebellum Texas," *Southwestern Historical Quarterly* 73 (April 1970): 431–55.

15. Ibid., 431–55.

16. Ibid., 431–55.

17. Winfrey, *Julien Sidney Devereux*, 42–43.

18. *Texas Almanac for 1857*, advertisements.

19. Olmsted, *Journey through Texas*, 64, 381–82.

20. Affleck, *Southern Rural Almanac*, 134.

21. Ibid., 20, 124.

22. Ibid., 64, 74.

23. Ibid., 102.

24. Curlee, "Study of Texas Slave Plantations," 212–18.

25. Affleck, *Southern Rural Almanac*, 11–16, 144.

26. Affleck, *Southern Rural Almanac*, 146–48; Curlee, "Study of Texas Slave Plantations," 336–40.

27. Sue Winton Moss, "Construction and Development of the Barrington Plantation," in *The Anson Jones Plantation: Archaeological and Historical Investigations at 41WT5 and 41WT6, Washington County, Texas*, 49–65; Sue Winton Moss, "A Plantation Model for Texas," in *Anson Jones Plantation*, 67–85.

28. Winfrey, *Julien Sidney Devereux*, 32–91; Abigail Curlee, "The History of a Texas Slave Plantation, 1831–63," *Southwestern Historical Quarterly* 26 (October 1922): 79–127.

29. Curlee, "History of a Texas Slave Plantation," 79–127; Marie Beth Jones, *Peach Point Plantation: The First 150 Years,* 105–25; Curlee, "Study of Texas Slave Plantations," 213; Kelley, "Plantation Frontiers," 86.

30. Elizabeth Silverthorne, *Plantation Life in Texas,* 28–36; J. S. Sydnor, "Auction Sales," in *Texas Almanac for 1860,* n.p.; Eugene C. Barker, "The African Slave Trade in Texas," *Quarterly of the Texas State Historical Association* 6 (October 1902): 145–58.

31. Annie Lee Williams, *A History of Wharton County, 1846–1961,* 105–7.

32. Silverthorne, *Plantation Life in Texas,* 36; Olmsted, *Journey through Texas,* 107, 237, 501.

33. Silverthorne, *Plantation Life in Texas,* 37; Williams, *History of Wharton County,* 109.

34. Silverthorne, *Plantation Life in Texas,* 39–41.

35. George P. Rawick, Jan Hillegas, and Ken Lawrence, eds., *The American Slave: A Composite Autobiography,* vol. 16, pt. 3, 166, 167, 178, 189, 206, 265, and pt. 4, 16, 43, 67, 76; Curlee, "Study of Texas Slave Plantations," 243–55.

36. Silverthorne, *Plantation Life in Texas,* 41–55; Barker, "African Slave Trade," 145–58.

37. Noah Smithwick, *The Evolution of a State or Recollections of Old Texas Days,* 24; Eric C. Caren, *Texas Extras: A Newspaper History of the Lone Star State, 1835–1935,* 18; Olmsted, *Journey through Texas,* 507–9.

38. B. A. Botkin, ed., *Lay My Burden Down,* 160–62.

39. Ibid., 159–60.

40. Ibid., 172–73.

41. Ibid., 75–77.

CHAPTER 8

1. Robin W. Doughty, *Wildlife and Man in Texas,* 44–78; Mary S. Helm, *Scraps of Early Texas History,* ed. and ann. Lorraine Jeter, 3, 26.

2. J. C. Clopper, "J. C. Clopper's Journal and Book of Memoranda for 1828," *Quarterly of the Texas State Historical Association* 13 (July 1909): 44–80; *A Visit to Texas,* 92.

3. John C. Duval, *Early Times in Texas; or, The Adventures of Jack Dobell,* 143, 240–41.

4. *Visit to Texas,* 93.

5. William Bollaert, *William Bollaert's Texas,* ed. William Eugene Hollon and Ruth Lapham Butler, 252–57; Daniel Shipman, *Frontier Life,* 59.

6. Noah Smithwick, *The Evolution of a State or Recollections of Old Texas Days,* 215–18.

7. Shipman, *Frontier Life,* 57–58.

8. Ibid., 58–59.

9. Bollaert, *William Bollaert's Texas,* 252–57.

10. Taylor J. Allen, *Early Pioneer Days in Texas,* 29–39; Roger W. Rodgers, "Candid Columns: Life as Revealed in Antebellum Newspaper Advertising in Northeast Texas," *East Texas Historical Association* 38, no. 2 (2000): 40–53.

11. Elise Waerenskjold, *The Lady with the Pen,* ed. C. A. Clausen, 29–30, 37.

12. Terry G. Jordan, *Trails to Texas,* 25–33, 36, 57, 91, 118–21.

13. Jordan, *Trails to Texas,* 25–124; Frederick Law Olmsted, *A Journey through Texas,* 364.

14. Donald E. Worcester, *The Texas Longhorn,* 18–19; Jordan, *Trails to Texas,* 77–81.

15. Bollaert, *William Bollaert's Texas,* 249–50; Andrew Forest Muir, ed., *Texas in 1837, an Anonymous Contemporary Narrative,* 66–67; Olmsted, *Journey through Texas,* 169.

16. *Visit to Texas,* 90–92, 116–23; J. M. Nance, ed., "A Letter Book of Joseph Eve, United States Chargé d'Affaires to Texas," *Southwestern Historical Quarterly* 43, no. 4 (April 1940): 488.

17. Nance, "Letter Book of Joseph Eve," 488.

18. Jordan, *Trails to Texas,* 83–102.

19. Ibid., 103–24.

20. Jordan, *Trails to Texas,* 51–58; Worcester, *Texas Longhorn,* 9–15.

21. Jordan, *Trails to Texas,* 72–73; Worcester, *Texas Longhorn,* 21–24; Muir, *Texas in 1837,* 65–67; Smithwick, *Evolution of a State,* 75.

22. Worcester, *Texas Longhorn,* 25–30; Viktor Bracht, *Texas in 1848,* trans. Charles Frank Schmidt, 53–54.

23. Armando C. Alonzo, *Tejano Legacy,* 79–80; Jean Louis Berlandier, *Journey to Mexico during the Years 1826 to 1834,* trans. Sheila M. Ohlendorf, Josette M. Bigelow, and Mary M. Standifer, 2:430, 542–44.

24. Andrés Tijerina, *Tejano Empire: Life on the South Texas Ranchos,* xxiii–xxx, 122–33; Alonzo, *Tejano Legacy,* 88–89.

25. Lowe and Campbell, 1987, 27–31; Jordan, *Trails to Texas,* 125–39.

26. Ferdinand Roemer, *Texas, with Particular Reference to German Immigration and the Physical Appearance of the Country,* trans. Oswald Mueller, 191; Olmsted, *Journey through Texas,* 261; Duval, *Early Times in Texas,* 143, 240–41.

27. Olmsted, *Journey through Texas,* 274.

28. Allen, *Early Pioneer Days,* 15–16; Waerenskjold, *Lady with the Pen,* 36.

29. Olmsted, *Journey through Texas,* 66, 86, 91.

30. Abigail Curlee, "A Study of Texas Slave Plantations, 1822–1865" (Ph.D. diss., University of Texas, 1932), 335; Thomas Affleck, *Affleck's Southern Rural Almanac, and Plantation and Garden Calendar for 1860,* 84–85, 92.

31. F. S. Wade, "The Mustang Hunt," in *Papers concerning Robertson's Colony in Texas,* ed. Malcolm D. McLean, 3:375–99.

32. George Wilkins Kendall, "Sheep Raising in Texas," in *Texas Almanac for 1859,* 125–26; Eric C. Caren, *Texas Extras: A Newspaper History of the Lone Star State, 1835–1935,* 43; Waerenskjold, *Lady with the Pen,* 36–37.

33. Kendall, "Sheep Raising in Texas," in *Texas Almanac for 1859,* 125–28; George Wilkins Kendall, "Sheep Raising in Texas," in *Texas Almanac for 1858,* 134–36.

34. Thomas Decrow, "Sheep Raising in Texas," in *Texas Almanac for 1859,* 128–29.

35. Olmsted, *Journey through Texas,* 258–59.

36. *Visit to Texas,* 198–99, 224–25, 80–81.

37. Allen, *Early Pioneer Days,* 77–78; B. A. Botkin, ed., *Lay My Burden Down,* 22.

CHAPTER 9

1. Ferdinand Roemer, *Texas, with Particular Reference to German Immigration and the Physical Appearance of the Country*, trans. Oswald Mueller, 6; J. W. Latimer, "The Wheat Region and Wheat Culture in Texas," in *Texas Almanac for 1859*, 64–71.

2. Thomas Affleck, *Affleck's Southern Rural Almanac, and Plantation and Garden Calendar for 1860*, 21, 32, 42; I. T. Tinsley, "Letter from I. T. Tinsley, of Brazoria Co.," in *Texas Almanac for 1859*, 76–77; Sue Winton Moss, "Construction and Development of the Barrington Plantation," in *The Anson Jones Plantation: Archaeological and Historical Investigations at 41WT5 and 41WT6, Washington County, Texas*, 49–65; Frederick Law Olmsted, *A Journey through Texas*, 133–34.

3. Moss, "Barrington Plantation," 49–65; Affleck, *Southern Rural Almanac*, 64, 74, 86, 94.

4. Affleck, *Southern Rural Almanac*, 104.

5. Ibid., 114, 124.

6. Olmsted, *Journey through Texas*, 94–95, 82, 136, 67.

7. Daniel Shipman, *Frontier Life*, 36–37.

8. Raymond E. White, "Cotton Ginning in Texas to 1861," *Southwestern Historical Quarterly* 61 (October 1957): 257–69.

9. White, "Cotton Ginning in Texas," 257–69; Karen Gernhardt Britton, *Bale O'Cotton: The Mechanical Art of Cotton Ginning*, 21–29.

10. White, "Cotton Ginning in Texas," 257–69; Britton, *Bale O'Cotton*, 21–29; *Texas Almanac for 1857*, advertisements; Olmsted, *Journey through Texas*, 206; *Texas Almanac for 1860*, advertisements.

11. Affleck, *Southern Rural Almanac*, 21, 29, 88,108–10.

12. Olmsted, *Journey through Texas*, 137; Tinsley, "Letter from I. T. Tinsley," 76–77; Affleck, *Southern Rural Almanac*, 32; Abigail Curlee, "A Study of Texas Slave Plantations, 1822–1865" (Ph.D. diss., University of Texas, 1932), 197–205.

13. Affleck, *Southern Rural Almanac*, 32, 42, 54, 64, 74, 114; *A Visit to Texas*, 134–35; Marie Beth Jones, *Peach Point Plantation: The First 150 Years*, 45; Curlee, "Study of Texas Slave Plantations," 197–205.

14. Moss, "Barrington Plantation," 49–65; Abigail Curlee, "The History of a Texas Slave Plantation, 1831–63," *Southwestern Historical Quarterly* 26 (October 1922): 79–127; Curlee, "Study of Texas Slave Plantations," 200–2.

15. Affleck, *Southern Rural Almanac*, 74–75.

16. Curlee, "Study of Texas Slave Plantations," 203–4.

17. William Russell Johnson, *A Short History of the Sugar Industry in Texas*, 10–13.

18. Jones, *Peach Point Plantation*, 45.

19. Johnson, *Sugar Industry in Texas*, 13–15.

20. Olmsted, *Journey through Texas*, 244–45; Johnson, *Sugar Industry in Texas*, 15–18.

21. Tinsley, "Letter from I. T. Tinsley," 76–77.

22. Tinsley, "Letter from I. T. Tinsley," 76–77; J. D. Waters, "Letter from Col. Waters, of Fort Bend Co.," in *Texas Almanac for 1859*, 77–80; Johnson, *Sugar Industry in Texas*, 21–22; Curlee, "History of a Texas Slave Plantation," 79–127; *Visit to Texas*, 92.

23. Johnson, *Sugar Industry in Texas,* 23–25; Waters, "Letter from Col. Waters," 77–80.

24. Curlee, "Study of Texas Slave Plantations," 186–97; Johnson, *Sugar Industry in Texas,* 26–30.

25. Johnson, *Sugar Industry in Texas,* 26–38.

26. *Texas Almanac for 1860,* "Staple Agricultural Crops of Texas," 217.

27. Latimer, "Wheat Region," 64–71.

28. Elise Waerenskjold, *The Lady with the Pen,* ed. C. A. Clausen, 28–29.

29. Latimer, "Wheat Region," 64–71.

30. Samuel Mather, "Remarks on the Cultivation of Wheat in Williamson and Adjoining Counties," in *Texas Almanac for 1859,* 74–76.

31. Latimer, "Wheat Region," 64–71.

32. Taylor J. Allen, *Early Pioneer Days in Texas,* 17.

33. Willard W. Cochrane, *The Development of American Agriculture: A Historical Analysis,* 67; Affleck, *Southern Rural Almanac,* 89.

34. Robert L. Ardrey, *American Agricultural Implements,* 5–126; Cochrane, *Development of American Agriculture,* 67–71; Affleck, *Southern Rural Almanac,* 30.

35. Noah Smithwick, *The Evolution of a State or Recollections of Old Texas Days,* 24.

36. Jonnie Lockhart Wallis, *Sixty Years on the Brazos: The Life and Letters of Dr. John Washington Lockhart, 1824–1900,* 195–97.

37. Smithwick, *Evolution of a State,* 225–27.

38. Affleck, *Southern Rural Almanac,* 99.

39. Ibid., 88, 112, reverse of title page.

40. Elizabeth Silverthorne, *Plantation Life in Texas,* 120–26.

41. Robert Hancock Hunter, *The Narrative of Robert Hancock Hunter,* 8.

42. Silverthorne, *Plantation Life in Texas,* 120–26; William Bollaert, *William Bollaert's Texas,* ed. William Eugene Hollon and Ruth Lapham Butler, 185, 261, 267.

43. Abigail Curlee Holbrook, "Cotton Marketing in Antebellum Texas," *Southwestern Historical Quarterly* 73 (April 1970): 431–55; Pamela Ashworth Puryear and Nathan Winfield, Jr., *Sandbars and Sternwheelers: Steam Navigation on the Brazos,* 47, 68; William Ransom Hogan, *The Texas Republic: A Social and Economic History,* 70; Olmsted, *Journey through Texas,* 90; Keith Guthrie, *Texas' Forgotten Ports,* 200; Roger W. Rodgers, "Candid Columns: Life as Revealed in Antebellum Newspaper Advertising in Northeast Texas," *East Texas Historical Association* 38, no. 2 (2000): 40–53; Eric C. Caren, *Texas Extras: A Newspaper History of the Lone Star State, 1835–1935,* 39; S. G. Reed, *A History of the Texas Railroads,* 27–28.

44. *Texas Almanac for 1860,* "Staple Agricultural Crops of Texas," 217–18.

45. Guthrie, *Texas' Forgotten Ports,* 1–80, 110–47, 170–88, 197–200; *Texas Almanac for 1860,* "Commerce of Indianola," 224.

46. Curlee, "History of Texas Slave Plantations," 209–10.

47. *Texas Almanac for 1860,* "Commerce of Galveston," 222.

48. *Texas Almanac for 1860,* "Commerce of Indianola," 224.

49. J. H. Davis, "Commerce of Port Lavaca," in *Texas Almanac for 1860,* 225.

50. *Texas Almanac for 1860,* "Commerce of Indianola," 224.

51. George P. Rawick, Jan Hillegas, and Ken Lawrence, eds., *The American Slave: A Composite Autobiography,* vol. 16, pt. 4, 47–49; Guthrie, *Texas' Forgotten Ports,* 198.

52. Wallis, *Sixty Years on the Brazos,* 198–99.

53. Bollaert, *William Bollaert's Texas,* 270.

54. Mrs. E. M. Houstoun, *Texas and the Gulf of Mexico; or, Yachting in the New World,* 230–31.

55. Olmsted, *Journey through Texas,* 90, 59–60.

56. Sean Michael Kelley, "Plantation Frontiers: Race, Ethnicity, and Family along the Brazos River of Texas, 1821–1886" (Ph.D. diss., University of Texas at Austin, 2000), 221–22.

57. Reed, *Texas Railroads,* 42–47.

58. Ibid., 42–47.

59. Reed, *Texas Railroads,* 42–47; *Texas Almanac for 1859,* 131–32; *Texas Almanac for 1860,* advertisements.

60. Reed, *Texas Railroads,* 50–84, 122–24.

BIBLIOGRAPHY

Affleck, Thomas. *Affleck's Southern Rural Almanac, and Plantation and Garden Calendar for 1860.* Brenham, Tex.: Privately printed, n.d., ca. 1859. A facsimile of the first edition. Brenham, Tex.: New Year's Creek Settlers Association, 1986.

Allen, Taylor J. *Early Pioneer Days in Texas.* Dallas: Wilkinson Printing, 1918.

Alonzo, Armando C. *Tejano Legacy.* Albuquerque: University of New Mexico Press, 1998.

Anderson, John Q. *Tales of Frontier Texas, 1830–1860.* Dallas: Southern Methodist University Press, 1966.

Ardrey, Robert L. *American Agricultural Implements.* Chicago: Privately printed, 1894. Reprint, New York: Arno Press, 1972.

Arneson, Edwin P. "Early Irrigation in Texas." *Southwestern Historical Quarterly* 25, no. 2 (October 1921): 121–30.

Barker, Eugene C. "The African Slave Trade in Texas." *Quarterly of the Texas State Historical Association* 6 (October 1902): 145–58.

Barker, Eugene C., ed. *History of Texas.* Dallas: Southwest Press, 1929.

Berlandier, Jean Louis. *The Indians of Texas in 1830.* Edited by John C. Ewers. Translated by Patricia Reading Leclercq. Washington, D.C.: Smithsonian Institution Press, 1969.

———. *Journey to Mexico during the Years 1826 to 1834.* Translated by Sheila M. Ohlendorf, Josette M. Bigelow, and Mary M. Standifer. Austin: Texas State Historical Association, 1980.

Béxar Archives Translations. Austin: University of Texas at Austin, 1983. Microfilm.

Bishko, Charles J. "The Peninsular Background of Latin American Cattle Ranching." *Hispanic American Historical Review* 32 (November 1952): 491–515.

Bishop, Curtis. *Lots of Land.* Austin: Steck, 1949.
Bollaert, William. *William Bollaert's Texas.* Edited by William Eugene Hollon and Ruth Lapham Butler. Norman: University of Oklahoma Press, 1956.
Bolton, Herbert E. *Texas in the Middle Eighteenth Century: Studies in Spanish Colonial History and Administration.* Austin: University of Texas Press, 1970.
Botkin, B. A., ed. *Lay My Burden Down.* Athens: University of Georgia Press, 1989.
Bracht, Viktor. *Texas in 1848.* Translated by Charles Frank Schmidt. San Antonio: Naylor, 1931. Reprint, Manchaca, Tex.: German-Texan Heritage Society, 1991.
Britton, Karen Gerhardt. *Bale O'Cotton: The Mechanical Art of Cotton Ginning.* College Station: Texas A&M University Press, 1992.
Bugbee, Lester G. "Slavery in Early Texas." *Political Science Quarterly* 13 (September 1898): 389–412.
Burke, James Wakefield. *Missions of Old Texas.* Cranbury, N. J.: Barnes, 1971.
Campbell, Randolph B. *An Empire for Slavery: The Peculiar Institution in Texas, 1821–1865.* Baton Rouge: Louisiana State University Press, 1989.
———. *Gone to Texas: A History of the Lone Star State.* New York: Oxford University Press, 2003.
Caren, Eric C. *Texas Extras: A Newspaper History of the Lone Star State, 1835–1935.* Edison, N.J.: Castle Books, 1999.
Carlson, Paul H. *Texas Woollybacks.* College Station: Texas A&M University Press, 1982.
Casañas de Jesus Maria, Fray Francisco. "Descriptions of the Tejas or Asinai Indians, 1691–1722. Part I. Fray Francisco Casañas de Jesús María to the Viceroy of Mexico." Translated by Mattie Austin Hatcher. *Southwestern Historical Quarterly* 30 (January 1927): 206–18.
Castañeda, Carlos E. *Our Catholic Heritage in Texas, 1519–1936.* 2 vols. Austin: Von Boeckmann-Jones, 1936–58.
Cat Spring Agricultural Society. *Century of Agricultural Progress: 1856–1956.* Cat Spring, Tex.: Cat Spring Agricultural Society, 1956.
Chevalier, François. *Land and Society in Colonial Mexico.* Translated by Alvin Eustis. Berkeley and Los Angeles: University of California Press, 1963.
Chipman, Donald E. *Spanish Texas, 1519–1821.* Austin: University of Texas Press, 1992.
Clopper, J. C. "J. C. Clopper's Journal and Book of Memoranda for 1828." *Quarterly of the Texas State Historical Association* 13 (July 1909): 44–80.
Cochrane, Willard W. *The Development of American Agriculture: A Historical Analysis.* 2d ed. Minneapolis: University of Minnesota Press, 1993.
Coleman, James M. *Aesculapius on the Colorado.* Austin: Encino Press, 1971.
Connor, Seymour V. "Log Cabins in Texas." *Southwestern Historical Quarterly* 53 (1949): 105–16.
Cross, F. M. *A Short Sketch-History from Personal Reminiscences of Early Days in Central Texas.* Belton: Bell County Historical Commission, 1979.
Curlee, Abigail. "The History of a Texas Slave Plantation, 1831–63." *Southwestern Historical Quarterly* 26 (October 1922): 79–127.
———. "A Study of Texas Slave Plantations, 1822–1865." Ph.D. diss., University of Texas, 1932.
Dabbs, J. Autrey, trans. *The Texas Missions in 1785.* Vol. 3, no. 6, *Preliminary Studies of the Texas Catholic Historical Society.* Austin: Texas Knights of Columbus Historical Commission, 1940.

Dary, David. *Cowboy Culture: A Saga of Five Centuries.* Lawrence: University Press of Kansas, 1981.

Davidson, Billy. "Preliminary Study of the Faunal Remains from the George C. Davis Site, 1968–70 Excavations." App. 2 in "1968–1970 Archeological Investigations at the George C. Davis Site, Cherokee County, Texas," by Dee Ann Story. *Bulletin of the Texas Archeological Society* 68 (1997): 108–13.

Davis, J. H. "Commerce of Port Lavaca." In *Texas Almanac for 1860,* 225. Galveston: Richardson, 1859.

Decrow, Thomas. "Sheep Raising in Texas." In *Texas Almanac for 1859,* 128–29. Galveston: Richardson, 1858.

De León, Arnoldo. *The Tejano Community, 1836–1900.* Albuquerque: University of New Mexico Press, 1982.

Dewees, W. B. *Letters from an Early Settler of Texas.* Waco: Texian Press, 1968.

Dobkins, Betty Eakle. *The Spanish Element in Texas Water Law.* Austin: University of Texas Press, 1959.

Domínguez, María Esther. *San Antonio, Tejas, en la época colonial (1718–1821).* Madrid: Ediciones de Cultura Hispánica, 1989.

Doolittle, William E. *Canal Irrigation in Prehistoric Mexico.* Austin: University of Texas Press, 1990.

Douay, Father Anastasius. "Narrative of La Salle's Attempt to Ascend the Mississippi in 1687." In *The Journeys of René Robert Cavelier, Sieur de La Salle,* edited by Isaac J. Cox, vol. 1. New York: A. S. Barnes, 1905.

Doughty, Robin W. *Wildlife and Man in Texas.* College Station: Texas A&M University Press, 1983.

Douglas F. H. "The Grass House of the Wichita and Caddo." Denver Art Museum, Department of Indian Art, Leaflet no. 42 (February 1932). Reprinted in *The Southern Caddo: An Anthology,* edited by H. F. Gregory. New York and London: Garland, 1986.

Dunn, William Edward. "Apache Relations in Texas, 1718–1750." *Quarterly of the Texas State Historical Association* 14 (1911): 198–274.

Duval, John C. *Early Times in Texas; or, The Adventures of Jack Dobell.* Lincoln: University of Nebraska Press, 1986.

Elvas, the Gentleman of. "The Narrative of the Expedition of Hernando de Soto." In *Spanish Explorers in the Southern United States, 1528–1543,* edited by Theodore H. Lewis. New York: Charles Scribner's Sons, 1907. Reprint, Austin: Texas State Historical Association, 1990.

Espinosa, Fray Isidro Felis de. "Descriptions of the Tejas or Asinai Indians, 1691–1722. Part IV. Fray Isidro Felis de Espinosa on the Asinai and Their Allies." Translated by Mattie Austin Hatcher. *Southwestern Historical Quarterly* 30 (October 1927): 150–80.

Fehrenbach, T. R. *Lone Star: A History of Texas and the Texans.* New York: Macmillan, 1968.

Fields, Ross C., Eloise F. Gadus, and L. Wayne Klement. "The Peerless Bottoms Site: A Late Caddoan Component at Cooper Lake, Hopkins County, Texas." *Bulletin of the Texas Archeological Society* 65 (1994): 55–114.

Flores, Dan L. *Jefferson and Southwestern Exploration.* Norman: University of Oklahoma Press, 1984.

Ford, Richard I. "Preliminary Report on the Plant Remains from the George C. Davis Site, Cherokee County, Texas, 1968–1970 Excavations." App. 1 in "1968–1970

Archeological Investigations at the George C. Davis Site, Cherokee County, Texas," by Dee Ann Story. *Bulletin of the Texas Archeological Society* 68 (1997): 104–107.

Forrestal, Peter P., and Paul J. Foik, eds. and trans. *The Solís Diary of 1767*. Vol. 1, no. 6, *Preliminary Studies of the Texas Catholic Historical Society.* Austin: Texas Knights of Columbus Historical Commission, 1931.

Foscue, Edwin J. "Agricultural History of the Lower Río Grande Valley Region." *Agricultural History* 8 (1934): 124–37.

Gates, Paul W. *The Farmer's Age: Agriculture, 1815–1860*. Vol. 3, *The Economic History of the United States*. New York: Holt, Rinehart and Winston, 1960.

Gleason, Mildred S. *Caddo: A Survey of the Caddo Indians in Northeast Texas and Marion County—1541–1840*. Jefferson, Tex.: Marion County Historical Commission, 1981.

Glick, Thomas F. *The Old World Background of the Irrigation System of San Antonio, Texas.* Southwestern Studies 35. El Paso: Texas Western Press, 1972.

Graham, Joe S. *El Rancho in South Texas: Continuity and Change from 1750.* Denton: University of North Texas Press, 1994.

———. "*Vaquero* Folk Arts and Crafts in South Texas." In *Hecho en Tejas,* edited by Joe S. Graham, 93–116. Denton: University of North Texas Press, 1991.

Guthrie, Keith. *Texas' Forgotten Ports.* Austin: Eakin Press, 1988.

Habig, Fray Marion A. *The Alamo Chain of Missions: A History of San Antonio's Five Old Missions.* Chicago: Franciscan Herald Press, 1968.

Harrison, James Christopher. "The Failure of Spain in East Texas: The Occupation and Abandonment of Nacogdoches, 1779–1821." Ph.D. diss., University of Nebraska-Lincoln, 1980.

Hatcher, M. A. *The Opening of Texas to Foreign Settlement, 1801–1821.* Bull. no. 2714. Austin: University of Texas, 1927.

Helm, Mary S. *Scraps of Early Texas History.* Edited and annotated by Lorraine Jeter. Austin, 1884. Reprint, Austin: Eakin Press, 1987.

Hogan, William Ransom. *The Texas Republic: A Social and Economic History.* Norman: University of Oklahoma Press, 1946. Reprint, Austin: University of Texas Press, 1969.

Holbrook, Abigail Curlee. "Cotton Marketing in Antebellum Texas." *Southwestern Historical Quarterly* 73 (April 1970): 431–55.

Holley, Mary Austin. *Texas.* Lexington, Ky.: J. Clarke, 1836. Reprint, Austin: Texas State Historical Association, 1990.

Houstoun, Mrs. E. M. *Texas and the Gulf of Mexico; or, Yachting in the New World.* Philadelphia: G. B. Zieber, 1845. Reprint, Austin: Steck-Warlick, 1968.

Hunter, Robert Hancock. *The Narrative of Robert Hancock Hunter.* Austin: Encino Press, 1966.

Hutchins, Wells A. "The Community Acequia: Its Origin and Development." *Southwestern Historical Quarterly* 31 (1927–28): 28.

Jackson, Jack. *Los mesteños.* College Station: Texas A&M University Press, 1986.

John, Elizabeth A. H. *Storms Brewed in Other Men's Worlds.* College Station: Texas A&M University Press, 1975.

Johnson, William Russell. *A Short History of the Sugar Industry in Texas.* Houston: Texas Gulf Coast Historical Association, 1961.

Jones, Marie Beth. *Peach Point Plantation: The First 150 Years.* Waco: Texian Press, 1982.

Jordan, Terry G. *German Seed in Texas Soil: Immigrant Farmers in Nineteenth-Century Texas.* Austin: University of Texas Press, 1966.

———. *North American Cattle-Ranching Frontiers.* Albuquerque: University of New Mexico Press, 1993.

———. *Texas Log Buildings: A Folk Architecture.* Austin: University of Texas Press, 1978.

———. *Trails to Texas.* Lincoln: University of Nebraska Press, 1981.

Joutel, Henri. "Historical Journal." In *The Journeys of René Robert Cavelier, Sieur de La Salle,* edited by Isaac Joslin Cox, vol. 2. New York: A. S. Barnes, 1905.

Kelley, Sean Michael. "Plantation Frontiers: Race, Ethnicity, and Family along the Brazos River of Texas, 1821–1886." Ph.D. diss., University of Texas at Austin, 2000.

Kendall, George Wilkins. "Sheep Raising in Texas." In *Texas Almanac for 1858,* 134–36. Galveston: Richardson, 1857.

———. "Sheep Raising in Texas." In *Texas Almanac for 1859,* 126–28. Galveston: Richardson, 1858.

Latimer, J. W. "The Wheat Region and Wheat Culture in Texas." In *Texas Almanac for 1859,* 64–71. Galveston: Richardson, 1858.

La Vere, David. *Life among the Texas Indians: The WPA Narratives.* College Station: Texas A&M University Press, 1998.

La Vere, David. *The Texas Indians.* College Station: Texas A&M University Press, 2004.

Lehmann, V. W. *Forgotten Legions: Sheep in the Rio Grande Plain of Texas.* El Paso: Texas Western Press, 1969.

Leutenegger, Benedict, ed. and trans. *Inventory of Mission Purissima Concepcion in the Province of Texas Made When the Fathers of Santa Cruz Gave It Up in 1772.* Archives of the College of Zacatecas, Old Spanish Missions Research Library, microfilm roll 4.5634–4.5664. San Antonio: Our Lady of the Lake University, 1772.

———. *Inventory of the Mission San Antonio de Valero: 1772.* Office of the State Archaeologist Special Report 23. Austin: Texas Historical Commission, 1977.

Lowe, Richard G., and Randolph B. Campbell. *Planters and Plain Folk: Agriculture in Antebellum Texas.* Dallas: Southern Methodist University Press, 1987.

Margry, Pierre, ed. *Découvertes et établissements des français dans l'ouest et dans le sud de l'Amérique Septentrionale, 1614–1754.* 6 vols. Paris: D. Jouaust, 1879–88.

Mather, Samuel. "Remarks on the Cultivation of Wheat in Williamson and Adjoining Counties." In *Texas Almanac for 1859,* 74–76. Galveston: Richardson, 1858.

McDonald, Archie P. *Hurrah for Texas! The Diary of Adolphus Sterne, 1838–1851.* Austin: Eakin Press, 1986.

Meyer, Michael C. *Water in the Hispanic Southwest.* Tucson: University of Arizona Press, 1984.

Montejano, David. *Anglos and Mexicans in the Making of Texas, 1836–1986.* Austin: University of Texas Press, 1987.

Morfí, Fray Juan Agustín. *Excerpts from the Memorias for the History of the Province of Texas.* Translated and edited by Frederick C. Chabot. San Antonio: Privately published, printed by Naylor Co., 1932.

———. *History of Texas: 1673–1779.* Translated by Carlos E. Castañeda. 2 vols. Albuquerque, N.M.: Quivira Society, 1935.

Morton, Ohland. *Terán and Texas: A Chapter in Texas-Mexican Relations.* Austin: Texas Historical Association, 1948.

Moss, Sue Winton. "Construction and Development of the Barrington Plantation." In *The Anson Jones Plantation: Archaeological and Historical Investigations at 41WT5 and 41WT6, Washington County, Texas,* 49–65. Reports of Investigations no. 2. College Station: Texas A&M University, Center for Environmental Archaeology, 1995.

———. "A Plantation Model for Texas." In *The Anson Jones Plantation: Archaeological and Historical Investigations at 41WT5 and 41WT6, Washington County, Texas,* 67–85. Reports of Investigations no. 2. College Station: Texas A&M University, Center for Environmental Archaeology, 1995.

Muir, Andrew Forest, ed. *Texas in 1837, an Anonymous Contemporary Narrative.* Austin: University of Texas Press, 1958.

Myers, Sandra L. *The Ranch in Spanish Texas: 1691–1800.* El Paso: Texas Western Press, 1969.

Nance, J. M., ed. "A Letter Book of Joseph Eve, United States Chargé d'Affaires to Texas." *Southwestern Historical Quarterly* 43, no. 4 (April 1940): 488.

Newkumet, Vynola Beaver, and Howard L. Meredith. *Hasinai: A Traditional History of the Caddo Confederacy.* College Station: Texas A&M University Press, 1988.

Olmsted, Frederick Law. *A Journey through Texas.* New York: Dix, Edwards, 1857. Reprint, Austin: University of Texas Press, 1978.

Pagés, Pierre Marie François de. *Travels round the World in the Years 1767, 1768, 1769, 1770, 1771.* 2 vols. London: J. Murray, 1791.

Perttula, Timothy K. "The Archeology of the Pineywoods and Post Oak Savanna of Northeast Texas." *Bulletin of the Texas Archeological Society* 66 (1995): 331–59.

———. *The Caddo Nation: Archaeological and Ethnohistoric Perspectives.* Austin: University of Texas Press, 1992.

Perttula, Timothy K., Myles R. Miller, Robert A. Ricklis, Daniel J. Prikryl, and Christopher Lintz. "Prehistoric and Historic Aboriginal Ceramics in Texas." *Bulletin of the Texas Archeological Society* 66 (1995): 175–235.

Perttula, Timothy K., Bob D. Skiles, and Bonnie C. Yates. "The Carlisle Site (41WD46), a Middle Caddoan Occupation on the Sabine River, Wood County, Texas." *Notes on Northeast Texas Archaeology* 1 (1993): 34–62.

Pike, Zebulon Montgomery. *The Journals of Zebulon Montgomery Pike: With Letters and Related Documents.* Edited by Donald Jackson. Norman: University of Oklahoma Press, 1906.

Pool, William C. *A Historical Atlas of Texas.* Austin: Encino Press, 1975.

Puryear, Pamela Ashworth, and Nathan Winfield, Jr. *Sandbars and Sternwheelers: Steam Navigation on the Brazos.* College Station: Texas A&M University Press, 1976.

Rawick, George P., Jan Hillegas, and Ken Lawrence, eds. *The American Slave: A Composite Autobiography.* Vol. 16, pt. 3 and 4. Westport, Conn.: Greenwood, 1972.

Reed, S. G. *A History of the Texas Railroads.* Houston: St. Clair, 1941.

Robertson, Felix. "Dr. Felix Robertson's Report." *The Weekly Messenger* (Russellville, Kentucky), June 17, 1826, 2, and *The Village Messenger* (Fayetteville, Tennessee), July 5, 1826, 2. Cited in *Papers concerning Robertson's Colony in Texas,* edited by Malcolm D. McLean, vol. 2, 601–6. Fort Worth: Texas Christian University Press, 1974.

Rodgers, Roger W. "Candid Columns: Life as Revealed in Antebellum Newspaper Advertising in Northeast Texas." *East Texas Historical Association* 38, no. 2 (2000): 40–53.

Roemer, Ferdinand. *Texas, with Particular Reference to German Immigration and the Physical Appearance of the Country.* Translated by Oswald Mueller. San Antonio: Standard Printing Company, 1935. Reprint, San Marcos, Tex.: German-Texan Heritage Society, 1983.

Rojas Rabiela, Teresa. "La agricultura en la época prehispánica." In *La agricultura en tierras mexicanas desde sus orígenes hasta nuestros días,* edited by Teresa Rojas Rabiela, 15–138. Miguel Hidalgo, Mexico: Editorial Grijalbo, 1990.

Romero Frizzi, María de los Ángeles. "La agricultura en la época colonial." In *La agricultura en tierras mexicanas desde sus orígenes hasta nuestros días,* edited by Teresa Rojas Rabiela, 139–215. Miguel Hidalgo, Mexico: Editorial Grijalbo, 1990.

Sáenz, Andrés. *Early Tejano Ranching: Daily Life at Ranches San José and El Fresnillo.* San Antonio: University of Texas Institute of Texan Cultures, 1999; Reprinted, College Station: Texas A&M University Press, 2001.

Schuetz, Mardith K. *Excavation of a Section of the Acequia Madre in Bexar County, Texas and Archeological Investigations at Mission San José in April 1968.* Austin: Texas Historical Survey Committee, Archeological Rep. no. 19. July 1970.

Scott, Florence Johnson. *Royal Land Grants North of the Río Grande.* Rio Grande City, Tex.: La Retama Press, 1969.

Shipman, Daniel. *Frontier Life: 58 Years in Texas.* Privately printed, 1879. Reprint, Pasadena, Tex.: Abbotsford, 1965.

Silverthorne, Elizabeth. *Plantation Life in Texas.* College Station: Texas A&M University Press, 1986.

Smithwick, Noah. *The Evolution of a State or Recollections of Old Texas Days.* Austin: Gammel Book Company, c. 1900. Reprint, Austin: University of Texas Press, 1983.

Strobel, Abner J. *The Old Plantations and Their Owners of Brazoria County, Texas.* Austin: Shelby, 1926. Reprinted in *Old Oyster Creek and Brazos Plantations.* Richmond, Tex.: Price/Ferguson, 1965.

Swanton, John R. *The Indians of the Southeastern United States.* Washington, D.C., and London: Smithsonian Institution Press, 1946.

———. *Source Material on the History and Ethnology of the Caddo Indians.* Washington, D.C.: Bureau of American Ethnology Bulletin 132, 1942.

Sydnor, J. S. "Auction Sales." In *Texas Almanac for 1860,* n.p. Galveston: Richardson, 1859.

Teja, Jesús F. de la. *San Antonio de Béxar: A Community on New Spain's Northern Frontier.* Albuquerque: University of New Mexico Press, 1995.

———. "A Fine Country with Broad Plains—the Most Beautiful in New Spain." In *On the Border: An Environmental History of San Antonio,* edited by Char Miller, 41–55. Pittsburgh: University of Pittsburgh Press, 2001.

Teja, Jesús F. de la, and John Wheat. "Béxar: Profile of a Tejano Community, 1820–1832." In *Tejano Origins in Eighteenth Century San Antonio,* edited by Gerald E. Poyo and Gilberto M. Hinojosa, 1–24. Austin: University of Texas Press, 1951.

Texas Almanac for 1857. Advertisements in *Texas Almanac for 1857,* unnumbered pages. Galveston: Richardson, 1856.

Texas Almanac for 1857. "Report of November 1855." In *Texas Almanac for 1857,* 37—38. Galveston: Richardson, 1856.

Texas Almanac for 1858. "African Slavery." In *Texas Almanac for 1858,* 132–33. Galveston: Richardson, 1857.

Texas Almanac for 1860. Advertisements in *Texas Almanac for 1860,* unnumbered pages. Galveston: Richardson, 1859.

Texas Almanac for 1860. "Commerce of Galveston." In *Texas Almanac for 1860,* 222. Galveston: Richardson, 1859.

Texas Almanac for 1860. "Commerce of Indianola." In *Texas Almanac for 1860,* 224. Galveston: Richardson, 1859.

Texas Almanac for 1860. "Staple Agricultural Crops of Texas." In *Texas Almanac for 1860,* 216–18. Galveston: Richardson, 1859.

Texas General Land Office. *Guide to Spanish and Mexican Land Grants in South Texas.* Austin: Texas General Land Office, 1999.

Thonhoff, Robert H. *El Fuerte del Cíbolo.* Austin: Eakin Press, 1992.

———. *The Texas Connection with the American Revolution.* Austin: Eakin Press, 1981.

Tijerina, Andrés. *Tejanos and Texas under the Mexican Flag, 1821–1836.* College Station: Texas A&M University Press, 1994.

———. *Tejano Empire: Life on the South Texas Ranchos.* College Station: Texas A&M University Press, 1998.

Tinsley, I. T. "Letter from I. T. Tinsley, of Brazoria Co." In *Texas Almanac for 1859,* 76–77. Galveston: Richardson, 1858.

Vielé, Teresa Griffin. *Following the Drum.* New York: Rudd & Carleton, 1858. Reprint, Lincoln: University of Nebraska, 1984.

A Visit to Texas. New York: Goodrich & Wiley, 1834. Facsimile reproduction. Austin: Steck, 1952.

Wade, F. S. "The Mustang Hunt." In *Papers concerning Robertson's Colony in Texas,* edited by Malcolm D. McLean, vol. 3, 375–79. Fort Worth: Texas Christian University Press, 1958.

Waerenskjold, Elise. *The Lady with the Pen.* Edited by C. A. Clausen. Waco, Tex.: Norwegian-American Historical Association, 1961. Reprint, Clifton, Tex.: Bosque Memorial Museum, 1979.

Wallis, Jonnie Lockhart. *Sixty Years on the Brazos: The Life and Letters of Dr. John Washington Lockhart, 1824–1900.* Los Angeles: Privately printed, 1930. Reprint, Waco: Texian Press, 1967.

Walters, Mark, and Patti Haskins. "Archaeological Investigations at the Redwine Site (41SM193), Smith County, Texas." *Journal of Northeast Texas Archaeology* 11 (1998): 1–38.

Waters, J. D. "Letter from Col. Waters, of Fort Bend Co." In *Texas Almanac for 1859,* 77–80. Galveston: Richardson, 1858.

Weddle, Robert S., and Robert H. Thonhoff. *Drama and Conflict: The Texas Saga of 1776.* Austin: Madrona Press, 1976.

White, Raymond E. "Cotton Ginning in Texas to 1861." *Southwestern Historical Quarterly* 61 (October 1957): 257–69.

Wilbarger, J. W. *Indian Depredations in Texas.* Austin: Hutchings Printing House, 1889. Reprint, Austin: Eakin Press and Statehouse Press, 1985.

Williams, Annie Lee. *A History of Wharton County, 1846–1961.* Austin: Von Boeckmann-Jones, 1972.

Winfrey, Dorman H. *Julien Sidney Devereux and His Monte Verdi Plantation.* Waco, Tex.: Texian Press, 1964.

Worcester, Don. *The Spanish Mustang.* El Paso: Texas Western Press, 1986.

———. *The Texas Longhorn.* College Station: Texas A&M University Press, 1987.

INDEX